Six and Four

The Complete Guide to 50 & 70MHz Amateur Radio

Don Field, G3XTT

Radio Society of Great Britain

Published by the Radio Society of Great Britain, 3 Abbey Court, Fraser Road, Priory Business Park, Bedford MK44 3WH, England.

ISBN: 9781 9050 8690 3

Publisher's note
The opinions expressed in this book are those of the author and not necessarily those of the RSGB. While the information presented is believed to be correct, the author, the publisher and their agents cannot accept responsibility for consequences arising from any inaccuracies or omissions.

Cover design: Kevin Williams
Sub-editing, layout and design: Steve Telenius-Lowe, 9M6DXX
Production: Mark Allgar, M1MPA

Printed in Great Britain

Contents

Preface

SOME FIVE YEARS AGO I embarked on a project to write a 6m handbook. The project was the brainchild of Kerry Rochester, G0LCS (now G8VR), a successful and enthusiastic 6m DXer, but, sadly, Kerry had to pull out due to health problems. At that point I took it on as I had previously been responsible for a major update of the *RSGB Operating Manual*, so had the experience of putting together a book of this sort. I had also been active on 6m for many years, though would by no means class myself alongside the sort of single-band enthusiasts that 6m attracts. The result was the *6 Metre Handbook*, published by the RSGB in 2008.

The response to that book in the intervening years has been very positive. There have been several reprints and I am extremely grateful for all the feedback I have received from readers around the world. But time moves on and it didn't make sense to reprint it again without an in-depth look at what might have changed during that time.

One very significant change was the emergence of the 4m band as a force in its own right. What had been largely a UK band is now a European and, increasingly, international band, with new countries gaining temporary or permanent access on a regular basis. To some extent I had anticipated these developments in the *6 Metre Handbook*, with a chapter dedicated to 4m. But it is time to move on – this book, while drawing heavily on the previous material, is positioned as a handbook for both bands, with almost equal emphasis on each. It is recognised that some readers will not have 4m privileges in their country but a lot of the advice here is, in any case, equally relevant to both bands.

So what else is new compared with that earlier *6 Metre Handbook*? I was hoping that, by now, there would be lots to say about all that exciting F2 propagation we had been enjoying, as during the last two solar maxima. Sadly (so far at least) that's not the case. But there is plenty of new equipment that has become available in the intervening years, especially by way of software-defined radios. And some significant advances have been made in antenna design. EME activity has increased (perhaps to compensate for the lack of F2 propagation). Capabilities for remote operation have come on apace. And there seem to be many more ways to stay abreast of band openings and activity wherever you are, through

Smartphones and all the other technology with which we surround ourselves nowadays.

My first exposure to 6m was on a visit to the MIT club station (W1MX) back in 1972. There was a 6m contest running and plenty of activity. Back then we didn't have access to 6m in the UK, for reasons which are explained elsewhere in this book. Eventually UK stations were allowed to apply for special permits for the band (a limited number of such permits were available) and then full access came in 1986, albeit with some power and antenna restrictions.

My own first foray on to the 6m band was in March 1988, with an elderly FT-726R transceiver (about 20 watts) and a rather tired 5-element Tonna antenna which I had picked up cheap at a rally. My early QSOs were around the UK but then, in June, came my first Sporadic E QSO, with ZB2IQ, quickly followed by 9H, CT, PA, DL and LA. On 6 June 1988 I made my first QSO 'across the Pond': W1WHL at 2243UTC, followed by eight others that same night (including VE1YX and the late W3ZZ, two of the stalwarts of 50MHz). These late night (European-time) transatlantic openings in June and July are a regular feature of 6m and one of the joys of operating the band – they seem to be just as regular at sunspot minimum as at sunspot peak.

The excitement was mounting at that time, with stations in countries awaiting 6m permits getting their share of the action by working crossband. In June of 1988 I see that I worked DL, YO and 4X, for example, all of whom were on 28MHz and responding to my 6m CQs. Most of the 10m activity was around 28885kHz, which had become the meeting point for 6m enthusiasts, the spot where information was shared and DX stations showed up while waiting for the MUF to rise. The demise of 28885kHz for this purpose came some years later when the *PacketCluster* system effectively made it redundant, though there probably isn't the camaraderie on the *Cluster* system (and on chat rooms such as the ON4KST facility) that seemed to exist on '885.

Back in those halcyon days there was plenty to be worked, even with a modest station – I see in January 1989 I was again working across the Pond and in February 1989 worked ZS and J52US (Dave Heil, K8MN) as well as more US stations. And so it went on. 6m enthusiasm was growing around the world as more countries started to gain access. US operators were enjoying the opportunity to work DX outside the US, and there was great excitement among the 6m fraternity each time a new one appeared. No *Cluster* then, so the telephone lines were kept busy, with serious competition to be among the first to achieve DXCC on the band. For those of us in northern Europe, life was easier than nowadays, because there wasn't so much competition from better-located stations in southern Europe (neither Spain nor Italy had 6m access at that stage). There are those who would also argue that the Cycle 22 peak was better for DX working than Cycle 23. Certainly, by that next peak I was finding life harder, possibly because propagation wasn't as good but also due to there being many more callers whenever a rare one appeared – so it was time to upgrade my station with a linear amplifier and a better antenna!

During those years Six enthusiasts developed a strong sense of being part of something special. They met up at social functions and swapped war stories (often about the 'one that got away' during that fleeting opening while they were out doing the weekly shop!) And when they did meet they all recognised each others' calls because it was always the same folk in the pile-up and, often, the same guy they had worked at the distant end. Callsigns like CO2KK, HH7PV or HC5K were familiar to 6m operators as the only stations active on the band from their respective countries. There was a real challenge in working some of those new ones, too. Countries like HB9, initially, had access only in the wee small hours, when there was no TV. QSOs often had to be made by meteor-scatter as there was little or no likelihood of Sporadic E at that time of night.

Things have moved on since those days, though readers who were there and have the tee-shirt will, I'm sure, recognise the atmosphere that I have described and perhaps be taken to reminisce about the 'good old days'. But 6m is still a challenge and a lot of fun as this book sets out to show. Nowadays there are new challenges, often made possible by technological developments such as the WSJT software suite. There is also more activity, with most countries now having a 6m allocation and most HF transceivers having 6m in their coverage. Some HF antennas even include 6m, the popular SteppIR range of HF Yagis, for example, having 6m as an add-on option. In response, most major expeditions now cater for 6m, especially if they are scheduled at a time of the year when Sporadic E or TEP propagation are likely. More and more are even catering for 6m EME. And the inclusion of 6m in the ARRL DXCC Challenge listings has also encouraged the more competitive HF DX chasers on to the band. So, all in all, a virtuous circle which makes for more activity and gives us more insights into 6m propagation as the years go by.

As for the 4m band, my experience goes back many years earlier than 6m, simply because 4m was very much part of the UK scene when I was first licensed in 1968, long before 6m became available to us. It was a band which I always enjoyed as part of the RSGB's VHF Field Day events and, from home, I was for a time the only 4m station active from Northamptonshire, so was popular with those who chased counties on the band. When 6m came along on a permanent basis I rather abandoned 4m although I used it for *PacketCluster* access for a while in the 1990s and, more recently, have been able to participate in my local radio club net on 4m FM by way of a converted PMR set. But in the past couple of years I have returned to DXing and contesting on the band with an FT-847 and small Yagi and have enjoyed making plenty of QSOs around Europe during the Sporadic E season. Like many of you, no doubt, I aim to build on that in the years to come.

There are some who worry that 4m will become just like any other band, losing its unique identity. Certainly, it has always enjoyed something of a special status with many regular users getting to know each other as friends. And, because there was relatively little commercial equipment available, it is a band that has encouraged homebrew construction. Balanced against this, though, there are new opportunities to explore propagation at that frequency and, in any case, who would want to deny those who are gaining the use of the band for the first time the enjoyment they are going to have.

Amateur radio operators have traditionally grouped their frequency bands according to their characteristics for communication. The low frequency bands, good for night-time propagation but suffering from D-layer absorption during the day. The high-frequency bands, enjoying the benefits of daytime energisation of the ionosphere by the sun to support long-distance propagation. The VHF bands, essentially line-of-sight except under occasional situations of enhancement. The microwave bands, requiring specialist equipment and with yet another set of characteristics. Like all such groupings, life is never quite as simple. But there is one amateur band that is almost unique in the way it behaves, sometimes acting as an HF band, with world-wide propagation, at other times acting much more as a VHF band, enjoying the benefits of Sporadic E, meteor scatter and other occasional propagation modes.

Because it has so many facets, 6m is both a challenge and an enigma and draws amateurs from both the VHF and HF worlds, especially those who want to try something new and different. 6m enthusiasts quickly become passionate about what they describe as 'The Magic Band' and it is hard to imagine any other waveband spawning its own club in the way that the UK Six Metre Group and SMIRK have arisen to cater for interests of 6m operators.

The *6 Metre Handbook* came about for exactly the same reason. Who would imagine a book specifically about 15m or even 2m? But 6m is sufficiently varied in what it has to offer, sufficiently unique in its characteristics, that it has a following all of its own. Increasingly, the same could be said of 4m, albeit for slightly different reasons at this point in time, mainly that it is new to many and therefore an opportunity to learn, experiment and chase some new ones.

One of the difficulties in writing a book such as this is that the audience varies from those who have never been on Six or Four and want a fairly detailed guide, to those who have 200+ countries (on 6m, at least) and could well teach the author a thing or two. The former want some basic advice, the latter perhaps want to reminisce a little. Equally, some readers will have an extensive amateur radio library covering the more general issues of propagation, operating, QSLing and so on, whereas others may be looking to these pages to give them some basics. So what I have tried to do is to start with the basics which, for some of you, may be old hat, so please be indulgent. For those who want to take these aspects further, there are plenty of useful references, many of which are listed in the bibliography, for example the ARRL and RSGB operating manuals, the ARRL and RSGB handbooks for more technical information, *The ARRL Antenna Book*, the *RSGB VHF / UHF Handbook* and a host of specialist propagation texts. Beyond the basics, the book covers facets of both bands such as history, a look at the 6m band from the perspective of users in different parts of the world, use of weak-signal modes and much more. Hopefully something for everyone.

That said, while almost all the feedback I received on the *6 Metre Handbook* was positive, which is heartening, the one minor criticism I got from a couple of people was that the book didn't feature circuit diagrams and constructional details for 6m equipment. This was quite deliberate and I make no apology. I see this very much as a resource book to motivate, suggest ideas for improvement, fill in some historical background and so on. But constructional material is readily available elsewhere, in RSGB, ARRL and other publications as well as on the Internet. And, to be frank, and although we may wish it otherwise, probably 90% of those who turn to 6m and even 4m nowadays will be using ready-made equipment, their interest primarily being operating rather than construction.

I must thank those who have made the book possible and I apologise in advance for any I have missed. The list below includes not only those who contributed to the original *6 Metre Handbook* (and most of their contributions have carried over into this edition), but also a number of new contributors, especially with respect to the additional 4m material. I am indebted to one and all.

I have relied heavily on support from the UKSMG, Kerry G8VR (previously G0LCS), and a number of 6m specialists such as Lance, W7GJ (renowned for his EME work on 6m). But many others have contributed by way of photos, reminiscences, the loan of historic material, input on specific topics, and so on. The list includes EY8MM, G0CHE, G0JHC, G0KSC, G0LCS, G1ZJP, the late G3FPQ, G3HGE, G3NGX, G3SED, G3SJX, G3WOS, G3ZYY, G4AFJ, G4BLH, G4CCZ, G4IGO, G6TGO, G8BCG, GM3SEK, GM4AFF, JA1RJU, JE1BMJ, K1JT, M5BXB, OZ2M, PP5XX / PY5CC, SV1DH, VK3OT / VK3SIX, the late W3ZZ, W6JKV, W7GJ, ZL3NW, ZL3TY, ZS6EZ, the UKSMG Committee as a whole, and the staff at RSGB who handled the actual publication including Mark Allgar, M3MPA; Kevin Williams, who was responsible for cover design, and Steve Telenius-Lowe, 9M6DXX / G4JVG, who did the final editing and prepared the book for printing. Without their help it would simply never have happened.

But, once again, this is very much a 'work in progress' and I would hope that any future edition can be made even more comprehensive than this one, building on this foundation. As such, I welcome all feedback. The usual caveats definitely apply. Any good material is probably due to input from one of the preceding, any errors are mine alone. I am more than happy to receive suggestions and input via e-mail (don@g3xtt.com).

I cannot end this preface without making mention of the late Gene Zimmerman, W3ZZ, who was kind enough to proofread the whole of the *6 Metre Handbook* and to contribute its Foreword. It is to his memory, as a great VHF (and especially 6m) enthusiast, that this edition is dedicated.

Don Field, G3XTT,
May 2013

1 Introduction

O VER THE PAST 20 years or so, the popularity of the 6m band has in creased dramatically, fuelled largely by new countries gaining access to the band. This is mainly due to that part of the spectrum being released from TV broadcasting, which was its principal use in many parts of the world before UHF television came along.

That popularity was also helped by the excellent propagation enjoyed on the band during the peak of cycle 23, around the turn of the Millennium. Long-distance contacts became almost commonplace and many 6m DXers saw their country totals creeping up towards, or even exceeding, the 200 mark. Sadly, the peak of cycle 24 has been less favourable to 50MHz propagation.

A third factor has been the inclusion of 6m in the ARRL DXCC Challenge, which has encouraged many HF DXers to 'have a go' on 6m for the first time.

In recent years the 4m (70MHz) band has also moved from being limited to just a handful of countries to now being available in many European countries and in a few locations elsewhere. This trend will certainly continue as that part of the spectrum becomes free of other activity. At the same time, the major manufacturers have spotted the trend and some are now adding the 4m band to at least some of their products.

Six News, the excellent quarterly journal of the UK Six Metre Group.

The author (centre) compares notes with Mike, G3SED (left), and Peter, G8BCG (ex-H44PT).

This book therefore aims to be an introduction to both bands, building on the original *6 Metre Handbook*, published by the RSGB in 2008 and which has proved to be very popular throughout the world, perhaps because there was a very real gap in the literature.

In these pages you will learn something of the history of both bands, be introduced to some of the key players, learn how to set up a station and how to operate on these fascinating bands. There is no way this book can be the ultimate resource, but hopefully it can point you in the right direction, for example at the wealth of information available nowadays via the Internet. Reader feedback is not only welcome, but encouraged, so that later editions can build on what appears here. It is hoped that there should be something here for readers in all parts of the world.

THE MAGIC OF 6M

Not for nothing is 6m often referred to as the 'Magic Band'. If you are a TV broadcaster, think what a nightmare it is to be using part of the spectrum where propagation is line of sight for much of the time but then, especially at certain times of the year, the signals travel much farther, resulting in mutual interference between stations that normally are well clear of one another. Those of us of a certain age will recall the days of VHF television when, especially in the summer months, our regular TV station suffered severe interference from stations in adjacent states or countries, often described to the uninitiated as "continental interference". What a disaster!

But for amateur radio operators, this sort of variability is exactly what makes the hobby so interesting. You might have been listening to a quiet band on and off for weeks, maybe just holding regular schedules with a few friends in the neighbourhood. Then you turn on one day and the band is full of signals, many of the stations speaking in other languages or with strange accents, using prefixes that you normally only hear on the HF bands. And yet, even in the quiet times, you are aware that some of the more serious enthusiasts are still managing to make long distance contacts, bouncing their signals off meteor trails or even off the moon. And not always on the traditional modes of SSB and CW either, but often using the latest signal processing software on their PCs to decode data signals quite inaudible to the human ear.

For many years, the American Radio Relay League (ARRL) held off introducing a version of the DX Century Club certificate for 6m, thinking that it would be demotivating to have an award that was impossible to achieve. Hard to believe this is the same band where quite a few serious DXers now have country scores of well over 200!

Six-metre and, increasingly, four-metre enthusiasts are a special breed. Not for them the relative predictability of 20m. They are prepared to listen to white noise for hours on end in the hope of maybe a 30-second window of opportunity where a DX signal rises above the noise just long enough to allow a contact to be made. The resulting sense of achievement really sets the adrenaline flowing, and is undoubtedly what makes these bands so popular. Clubs such as SMIRK (Six Meter International Radio Klub) and the UK Six Metre Group have much larger membership numbers than many similar organisations, reflecting the camaraderie between those who enjoy the special appeal of Six. In some ways the appeal is similar to that of 160m (Topband), where dedication, combined with an above-average level of technical knowledge and propagation expertise are required to reach the higher echelons of achievement. Those who populate the band are simply not satisfied with chatting through

JW ON SIX!

(This short piece, by Chris Gare, G3WOS, gives a flavour of the excitement when a 'new' one shows – most 6m enthusiasts will have similar stories of their own.)

On the evening of 10 July 1996 at around 2030UTC the JW beacon started coming into the UK at some considerable strength. For over half an hour it varied between S4 and S7. Neil, G0JHC, rang Matt, JW5NM, only to discover that he was not at home. Later Chris, G3WOS, in desperation rang Ole, JW8GV, who then said he would go to the shack. He said he was not too optimistic about getting the station on the air for some reason. Neil explained to me later that they needed to disconnect the beacon antenna and connect it to an HF transceiver / transverter arrangement. After ten minutes, which seemed to be an eternity, the beacon stopped keying for a few seconds and then restarted. Are they in the shack? A couple of minutes later the beacon shuts off! They *are* there! A quick call to Neil revealed that they had already faded out in the north of England, much to his annoyance.

Tuning the band suddenly reveals Peter, JW7QIA, calling CQ on 50.090MHz. I called but there was no response - wasn't he hearing me? This happened about six times, then suddenly I get a "G3W?". It was then but a few minutes to complete a QSO and get a "TU". Mayhem then broke out and at least G3FPQ and G3IBI got into his log. The QSOs took place at around 2120UTC. As it turns out, Peter was only visiting the island and working a few stations while there.

their local repeater on 2m, or having reliable long-distance propagation on the major HF bands. They want a *real* challenge. To get a flavour of the intensity with which 6m enthusiasts follow their particular aspect of the hobby, take a look at some of their dedicated websites, good examples being those of M5BXB, G3WOS, JA1RJU and the like. Some, like VK3SIX and K1SIX have even selected callsigns that advertise their lifelong love of 6m.

SOME HISTORY

Before 6m became available to the amateur radio community, amateurs in some countries were able to use the 56MHz (5m) band, a convenient frequency harmonically related to the lower bands (3.5, 7, 14, 28, 56MHz – no 21MHz back then). In the USA the 5m band became available in 1924, along with the 80, 40 and 20m bands (10m was authorised in 1927). The October 1924 issue of *QST* featured a 5-metre oscillator design, probably the first VHF constructional article to appear in print in an amateur radio publication. In the June 1929 issue of the UK magazine *Wireless World*, G6TW was reported to have won a prize, offered in the April issue of the magazine, for the first communication over 10 miles on 5 metres (presumably the first in the UK, given that 5 metres was already five years old in the USA).

The 13 December 1935 issue of *Wireless World* carried an article by a D R Parsons entitled 'Range of Five Metre Transmissions', in which he postulates ways in which 5-metre signals might propagate beyond the "theoretical" limit (at that time it was thought that the ionosphere was unable to support reflections at such "ultra-short" wavelengths). But it was in just the following year, 1936, that G5BY was the first European station to span the Atlantic on 56MHz, when his signals were heard by W2HXD, although no two-way contact ensued.

By 1938 the 56MHz distance record was held by W1EYM and W6DNS, a 2500-mile path. With the advent of war in 1939 activity on all bands was much reduced, with many countries closing down amateur radio activities completely. Technological developments during the war years had made it cost-effective to roll out TV broadcasting on VHF frequencies that, prior to the war, were considered to be of little use. Thus, immediately post-war there was quite an administrative battle over the use of the VHF spectrum.

Interestingly, the then ARRL General Manager, Kenneth Warner, during his 1944 negotiations with the FCC, is quoted as having said, "*We have previously characterised the performance of this band to you as being erratic, unpredictable, unreliable and unexpected, a band where anything can and generally does happen, and we have explained that its very eccentricities give it a peculiar charm for us, though they make it singularly bad for regular service.*"

In the USA the outcome of the negotiations was that amateurs lost the 56MHz band, which now fell within the 54 – 60MHz spread of TV's channel 2 but managed, instead, to get an allocation between 50 and 54MHz (TV channel 1 was 44 to 50MHz). The new 6m band, as it was now known, became available in the USA in March 1946 and activity soon picked up, with the first transcontinental QSO taking place on 14 June, between W6OVK (CA) and W2BYM (NJ). Before the year was out 6m was available to amateurs in Canada, Australia and New Zealand. By 1947 there was further activity, especially from Central and South America and some limited authorisations became available in Europe (the Netherlands, with 100 watts, and the UK with 25 watts. UK operators were limited to certain times of the day to avoid interference to TV services and the authorisations were available until 30 April 1948, see below).

UK FIVE-METRE FIRSTS
(source: Short Wave Magazine, May 1947)

France	G2FA	/ F8NW	29 March 1936
Italy	G5MQ	/ I1IRA	2 July 1938
Holland	G2AO	/ PA0PN	17 August 1939
North Africa	G5BY	/ FA8B	24 June 1946
Switzerland	G5BY	/ HB9CD	22 August 1946
Sweden	G5TH	/ SM5FS	24 May 1947
Denmark	GM6KH	/ OZ7G	24 May 1947
Belgium	G6DH	/ ON4KN	25 May 1947
Czechoslovakia	C5DY	/ OK2MV	22 June 1947
Malta	G6LK	/ ZB1AB	30 June 1947
Gibraltar	G2XC	/ ZB2A	22 July 1947

Immediately post-war the UK retained a 5m allocation, at 58MHz, and a number of tests were conducted with the hope of achieving a cross-band 5m / 6m QSO. This didn't happen and the first trans-Atlantic QSOs were made cross-band 10m / 6m, simply because while these tests were taking place the MUF never reached as high as 58MHz. Hence a request was made by UK amateurs to the UK General Post Office for a temporary 6m permit. As it happens, the 5m band was only available in the UK for about three years before those frequencies, as in the US, were required by the emerging national TV broadcasting service. Things didn't look good. In an editorial in the May 1947 issue of *Short Wave Magazine*, Austin Forsyth, G6FO, looking ahead to the Atlantic City conference, at which frequency allocations were to be discussed and decided, wrote, "*To 5-metre operators, the lack of any suggestion between 29.7 and 168 mc is disastrous. A band, however narrow (250 kc would do), in the 50 or 60 mc region, allocated on an international basis, would be sufficient for our needs*". It didn't happen. Amateurs did gain the 2m band, but anything between 29.7 and 144MHz was left to national licensing authorities, with the result that, for something like the next 40 years, there was no internationally-harmonised band.

Nevertheless, UK amateurs continued to lobby for a lower VHF band to replace 5m and in 1956, seven years after 5m was taken away, were allocated 70.2 to 70.4MHz (or Mc/s as it would have been referred to at that time), the 4m band. Unfortunately this was available only in the UK and a limited number of UK overseas territories, so British amateurs continued to press for a 50MHz allocation but, as with much of the rest of Europe, this didn't materialise (other than by way of occasional special permits) until the TV broadcasters finally started to abandon VHF in favour of their newer and more effective UHF allocations (more effective because they offered enough bandwidth for higher definition and colour, but also because they got away from the vagaries of VHF propagation – all that dreadful Sporadic E and stuff, which was exactly what the amateurs were interested in).

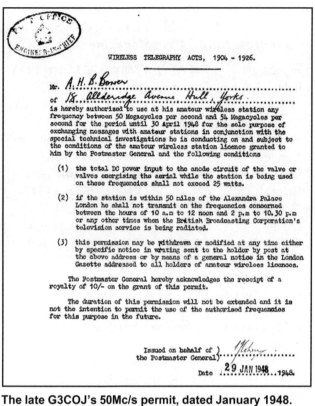

WIRELESS TELEGRAPHY ACTS, 1904 - 1926.

Mr. *A. H. B. Bower*

of *18. Alldridge Avenue Hull, Yorks.*

is hereby authorised to use at his amateur wireless station any frequency between 50 Megacycles per second and 54 Megacycles per second for the period until 30 April 1948 for the sole purpose of exchanging messages with amateur stations in conjunction with the special technical investigations he is conducting on and subject to the conditions of the amateur wireless station licence granted to him by the Postmaster General and the following conditions

(1) the total DC power input to the anode circuit of the valve or valves energising the aerial while the station is being used on these frequencies shall not exceed 25 watts.

(2) if the station is within 50 miles of the Alexandra Palace London he shall not transmit on the frequencies concerned between the hours of 10 a.m to 12 noon and 2 p.m to 10.30 p.m or any other times when the British Broadcasting Corporation's television service is being radiated.

(3) this permission may be withdrawn or modified at any time either by specific notice in writing sent to the holder by post at the above address or by means of a general notice in the London Gazette addressed to all holders of amateur wireless licences.

The Postmaster General hereby acknowledges the receipt of a royalty of 10/- on the grant of this permit.

The duration of this permission will not be extended and it is not the intention to permit the use of the authorised frequencies for this purpose in the future.

Issued on behalf of) the Postmaster General)

Date ... 29 JAN 1948 ...1948.

The late G3COJ's 50Mc/s permit, dated January 1948.

Anyway, back to the 6m story. In the immediate post-war years J9AAK was active from Okinawa, and amateurs in other parts of the world without 6m privileges were busy looking for cross-band contacts (10m to 6m). The first trans-Atlantic QSOs were in spring 1947 and some other rare DX was worked from Europe including MD5KW (Suez, operated by G5KW) and SU1HF (Egypt). The following news prompted a separate 'news flash' to be included in the April 1947 issue of *Short Wave Magazine*, the only occasion I have seen this unusual step taken, so the news was clearly considered highly memorable, "*At 1330 on March 26, PA0UN was received on 50 mc by ZS1AX, ZS1P and ZS1T, at up to S9*". At this time activity was confined to AM and CW. SSB didn't really make inroads on to 6m until the mid-60s or so. Activity continued to increase and, in 1956, the distance record was extended to 12,000 miles with a QSO between LU9MA and JA6FR. Give that this is half the circumference of the world, it was already clear that 6m supported world-wide propagation when the sunspot cycle was co-operating (that cycle is remembered by those who were around at the time as being quite exceptional, as indeed it was). The main limitation to a high country score on 6m was lack of 'new ones' to work, with most of Europe, for example, being off-limits because that part of the spectrum was used for TV broadcast purposes. That said, some special authorisations came into being in 1957 on the back of International Geophysical Year, bringing activity to the band from CT2 (Azores, now CU), CM (Madeira, now CT3), CT (Portugal), Norway, Sweden and Poland.

It was not until the 1980s that things started to change in Europe. In the UK a proposal emerged in 1980 to allow a limited number of amateurs access to the band. This move was supported by the Home Office (the licensing authority at that time) but an objection was received from the BBC and the plan was abandoned. Two years later things had moved on. The UK Six Metre Group (UKSMG) had been founded by Steve Richardson, G4JCC, to promote interest in the band and Jim Sleight, G3OJI, a senior engineer at the BBC, helped to overcome objections from that organisation. Some three hundred UK amateurs applied for experimental licences for the band, to be whittled down to 46, the number that the Home

Office had agreed to issue. In December of 1982 the GB3SIX beacon was licensed for 24-hour operation and in January 1983 the first permits were issued, allowing operation from 1 February outside TV hours (those were the days when TV actually closed down overnight – hard to believe now). By the end of the year this restriction had been lifted and 24-hour operation was allowed and in February 1984 it was announced that a further 60 permits were to be issued, bringing the total to 100. Lots of activity ensued, despite limitations on ERP and antenna height, with plenty of the permit holders managing to work across the Pond. This occasionally led to some interesting situations, such as with my good friend Dennis Robinson, GJ3YHU, calling a W5 station on the band who refused to work him on the basis that GJ3YHU wasn't a 'proper' callsign (the W5 presumably never previously having heard any stations from outside the US).

From 1 February 1986 all UK Class A licensees were allowed access to 6m, albeit with an ERP limit of 100 watts on SSB (25W on CW)

1965 ad for a 6m receiver and transmitter, from Whippany Laboratories, Inc.

and a maximum antenna height of 20m, horizontally polarised only. The main fear was that UK amateurs using 6m could cause interference to TV in mainland Europe, particularly in Belgium and France. In reality no problems seemed to arise and Class B licensees were given access to the band from 1 June 1987. From then on the various limitations were incrementally relaxed as VHF TV was gradually closed down not only in the UK but across Europe.

Similar stories were being enacted in many other countries throughout this period, usually with some sort of special permit in the first instance, allowing the national licensing authority to withdraw the band should problems occur. Once it was clear that all was well restrictions would gradually be withdrawn. Nevertheless it took something like 25 years before all of Europe, for example, had 6m access (Hungary being the last major country to gain access

THE CREATION OF UKSMG

(abridged from a 2008 letter by Steve, G4JCC, after he had been made an Honorary Member of UKSMG in recognition of his efforts in founding the Group and contributing to its early success).

The autumn of 1979 produced some outstanding propagation on six metres and I hold a tape containing a long QSO by Bob, VE1YX, with another Canadian operator on Six, recorded at that time, together with some other QSOs. These openings generated a lot of interest in VHF propagation and I had many conversations with Harold Rose, G8NWF, soon to become G4JLH. One of the points that arose was that there was great interest in six metres during the solar maxima but interest waned and people went to other bands and pursuits as propagation declined. Around this time I learned, before it was public knowledge, that all low-band television was to be switched off. I took this up and my contact promised to keep me informed of developments. Later he told me that the BBC wished to retain all their TV frequencies which meant the whole band from circa 40MHz to 80MHz. This information was very confidential at the time so I could make no direct use of it without compromising my informant, who was not a ham nor was he interested in amateur radio. However, it was obvious that there was an opportunity to regain a VHF band (now 6m) which we had lost to the BBC (this was the old 56Mc/s band; I had been a SWL on this frequency prior to WWII). So I decided to contact and have some discussions with amateurs who were or could be interested in a VHF band at 50MHz. I talked to G4JLH and, most importantly, G5KW with whom I had a very useful and productive meeting.

My discussions with Ken, G5KW, centred round the best course to pursue to retain the interest of radio amateurs. We were agreed that a few temporary licences and a beacon or two were not sufficient to retain this interest. Ken suggested that a column in *RadCom* about 6m matters might be the way to go or to form a branch of the 'Six Klub' which was an American organisation based in South Texas. So gradually our thoughts turned to having our own 6m club. I agreed to publicise this idea and gradually it grew but it needed a title, anything containing the word "Club" was out as it could be confused with the US one, so gradually the title UK Six Metre Group was adopted; then I began to write letters and solicit members from 6m operators (cross-band of course). The rest is history but it took a lot of discussion and work to reach this stage. The Merriman Committee was formed by government to decide what use should be made of the vacant TV frequencies and I, along with several other members, made representations to the Committee, which resulted in a recommendation for a band at 50MHz.

to the band, in 2005, though some, particularly European Russia, are currently inactive once again).

For a much more detailed account of 6m history, by far the best starting point is the series of articles written by Ken Willis, G8VR, for *Six News* (the UKSMG journal) and reproduced on the UKSMG website.

WHAT DOES SIX HAVE TO OFFER?

Six metres demonstrates the characteristics of both a VHF band and an HF band at different times. For much of the time the MUF is below 50MHz and propagation on Six is predominantly line-of-sight, as on the higher VHF bands. At such times the band can be handy for local 'ragchews', particularly if you don't have other VHF equipment in the shack but have an HF transceiver with 6m capability (which many do nowadays). There is no simple answer as to what you might expect the range to be, as it depends very much on your location and the type and height of antenna that you are using. But with a decent home-station set-up you might expect to be able to work similarly-equipped stations up to 100 miles (160km) or so without too much difficulty.

In many countries there is a network of FM repeaters to extend the workable range, exactly as on the other VHF bands. But don't expect much mobile activity on 6m; mobile antennas are much more manageable on 4m, 2m and 70cm (and 220MHz in North America). Some of these repeaters are connected to the Internet via systems such as *Echolink*, giving users access to the whole world via their 6m access channel. Up to date lists of repeaters can be found on the Internet. A list of UK 6m repeaters also appears at **Appendix C** to this book, current at the time of writing. A number of channels is also allocated to packet radio, for applications such as messaging, *PacketCluster*, APRS and Satgate. It is not the intention of this book to cover these specialist applications, which are well documented elsewhere and not specific to 6m, but the message is that 6m carries much the same sort of day-to-day activity as other VHF bands, applications which generally do not exist on the HF bands, largely due to bandwidth limitations.

But even when there is no 'skip' on 6m, whether from F layer, Sporadic E or other ionsospheric effects, you will hear stations working long distances by a variety of means. These are principally EME (Earth-Moon-Earth, also referred to as Moonbounce) and meteor-scatter. The names are self-explanatory and these propagation mechanisms are covered more fully elsewhere in this book. They help to keep alive interest in long-distance working on Six, even when the band is otherwise flat. EME requires a well-equipped station but meteor scatter is possible on almost a daily basis, even with a modestly-equipped station, particularly if you are using the appropriate WSJT software (again, covered later in this book). During the main meteor showers this mode becomes even more effective, permitting contacts up to something like 1400 miles (2300km).

For many 6m operators, though, the exciting times of the year, particularly when there is no F layer propagation, are those when Sporadic E ('E skip') propagation occurs. The mechanism is explained in Chapter 4 although even now there is some argument about exactly what causes Sporadic E. But from an operational point of view the key as-

The annual '6m BBQ' held at the home of Chris, G3WOS, is a good opportunity to meet fellow 6m enthusiasts and learn about current developments. During the 2011 BBQ Dennis, K7BV, spoke about his operations as 5J0BV from San Andres and Providencia .

pect is that Sporadic E occurs annually at very predicable times of the year. In the northern hemisphere the main Es season is in the May to August period but certain openings seem to be focused on an even more clearly-defined window in late June to early July, when, for example double and triple-hop Es support trans-Atlantic, coast-to-coast US and occasionally trans-Pacific contacts. As a result, many 6m DXpeditions are scheduled for this time of the year and serious 6m enthusiasts plan to take vacation time so that they can be on the band when openings occur. Signal strengths during Sporadic E openings can be high, so that even the most modest station can make successful contacts during these openings. The problems caused by Sporadic E (leading to mutual interference between stations hundreds of miles apart) was one of the reasons that TV broadcasters moved higher in frequency, though the move was also forced on them by the need for greater bandwidth to accommodate colour and higher definition broadcasts.

For operators in the lower latitudes (closer to the equator), for example in Southern Europe, trans-equatorial propagation (TEP) is also a reliable means of long-distance propagation for much of the year and this is covered elsewhere in this book.

Somewhat less predictable is auroral propagation, again described in the chapter on propagation. Auroras can occur at any time of year but they tend to be more common around the vernal and autumnal equinoxes. They follow major solar disturbances, so the keen 6m DXer will keep an eye not only on the bands (a strong aurora leads to a blackout of the HF bands) but also on the various space weather sites on the Internet.

The previous paragraphs have described how 6m can bring some interesting contacts year in and year out. But the fun really starts during years of high solar activity when the F2 MUF reaches 50MHz and 6m starts to behave like an HF band, with F layer propagation. When this happens world-wide contacts are possible. Sadly, Cycle 24 has not produced the sort of MUFs that were experienced during the last solar maximum, and it is easy to forget how Six can come to life at the peak of the cycle, as it did back around the turn of the Millennium. At that time European 6m operators were working regularly into the Far East and Australia, North American amateurs were seeing regular openings across the Atlantic, out into the Pacific and across their own continent. Occasionally some really long-haul contacts were made, such as the long-path contact between the late David Courtier-Dutton, G3FPQ, and KH6SX in Hawaii, at 0957UTC on 7 April 2002, a path of about 26,500km (and only worked once previously, between KH7R and G4EAT on 27 March 2000). These band openings can be fleeting and propagation will tend to peak at certain times of the year. In a sunspot maximum year F2 propagation begins in October, peaks in the month prior to the winter solstice and dies out sometime near the end of January. It is this type of propagation that allows high country totals to be achieved on 6m as it is only F layer propagation which will allow truly long-haul contacts on 6m, at least unless you are equipped for EME (but far fewer countries are likely to be activated on EME than will be active during F layer openings).

SOME 6M ACHIEVEMENTS

To put some of the foregoing in perspective, it is worth a short diversion to review some of the operating achievements on 6m to date. The first 6m DXCC awards were issued to K5FF and W5FF in January 1990. VE1YX was third and JA4MBM fourth to receive the award. The first European station to achieve DXCC on the band was G4AHN in 1991, receiving certificate number 29. I am indebted to Chris Burger, ZS6EZ, for compiling a comprehensive lists of early DXCCs on the band, which appears on his website [1].

At the time of writing (April 2013) the DXCC listings show an astonishing 48 amateurs with 200 or more entities credited on 6m (up from 22 in the previous edition of this book, despite the somewhat flat propagation in the intervening years). But what is perhaps most revealing is that the highest-placed North American station is K1TOL with 185. It is not surprising that W4DR, who used to lead the DXCC Challenge listings, is now in seventh place – it is obviously much tougher to achieve a high 6m country score from North America than from Europe or Japan. LZ2CC heads the 6m DXCC list with an astonishing 257, the highest UK entrant being Neil Carr, G0JHC, with 221. But one of the more interesting scores is that of Peter Sprengel, PY5CC (now PP5XX), also with a score of 221. It is several years since Peter was active on Six, but prior to that there was little if anything he had missed on the band. Peter's 200th country on Six was KH6ND/KH5 (Palmyra and Jarvis) in November 2000, making him almost certainly the first amateur to achieve 200 countries on the band. Peter lives in what must be the ideal location for 6m, positioned exactly right in respect of the geomagnetic equator to have reliable trans-equatorial propagation to most of the main areas of the globe and if he had been more active in recent years could possibly still be at the top of the list.

To put these scores in perspective, the UKSMG web page [2] shows 250 entities that have been worked from the UK of which two, FR/G and YV0, have not been accepted for DXCC. So G0JHC's 221 accredited is pretty impressive, particularly so as it has been achieved entirely by terrestrial means (no EME).

This is equally true of many of the others at the top of the 6m listings – very little has passed them by in terms of what could have been worked from their locality. The leading 6m operators really are a dedicated bunch! Similar lists of 'firsts' exist for many countries, so there could well be a list for your country if you search for it - a great starting point to know what is possible (though frustrating, perhaps, to see what was worked at the peak of the last cycle and to know that it may be some time before such conditions return!) Incidentally, and more about this later, but several remote entities which have been activated in recent years on 6m have only been workable on EME, so those who had no capability for that have missed out.

Most of these 200-plus scores were achieved with the help of a couple of sunspot peaks, but even at the bottom of the sunspot cycle it is astonishing what can be worked on the band. To give a flavour, UKSMG runs an annual 6m 'Marathon' lasting for three months over the main Sporadic E season. In 2007, from 6 May to 5 August, ON4IQ worked an astonishing 108 DXCC entities. The leading US participant, W1JJ, managed 78 (a much tougher proposition as European 6m operators have some 60 or so entities almost on their doorstep). It is not unusual in the summer from the UK to hear and work US and Caribbean stations on 6m while there is nothing doing on the high HF bands. 6m really is a unique band.

A QUICK TOUR OF 6M

The 6m band plan is covered in detail in Chapter 5. In broad terms, in Europe you will find beacons near the bottom of the band (quite deliberately, so that as the MUF rises past 50MHz the beacons should be the first signals to be heard, though in the US the beacon band is 50.060 to 50.080MHz and there have been suggestions that this allocation should be harmonised globally, if only to clear the bot-

COUNTRY	ALLOCATION	NOTES
Austria	50.000 – 52.000	100 watts, parts of OE1, OE3, OE4 can only transmit outside TV hours.
Azores	50.000 – 50.500	One year renewable licences, for propagation "tests" only. First Class licence required.
Belgium	50.000 – 52.000	6m available to ON amateurs only (not CEPT). 50W, but special higher-power permits may be available.
Canada	50.000 – 54.000	CEPT. 1000 watts (depending on licence class).
Cyprus	50.000 – 51.000	Available on a secondary (non-interference) basis.
Czech Republic	50.000 – 52.000	Czech Class A & B licensees only, 20 watts. Horizontal antennas only. No CEPT for 6m.
Denmark	50.000 – 52.000	1000 watts, depending on licence class.
Eire	50.000 – 52.000	20dBW, non-interference basis.
Estonia	50.000 – 52.000	(50.130 - 52.000MHz for Class D licensees).
France	50.200 – 51.200	5/100 watts EIRP (depending on location), no mobile operation, no CEPT.
Italy	50.000 – 51.000	500 watts (50 watts for Class B).
Malta	50.000 – 51.999	25 watts, 14dBW ERP, no mobile or portable.
Netherlands	50.000 – 52.000	120 watts PEP for A1A and J3E 50 – 50.450MHz, 30 watts PEP otherwise.
Norway	50.000 – 52.000	100 watts. JW and JX have same conditions.
Portugal	50.000 – 50.500	20 watts, valid for 5 years on non-interference basis.
UK	50.000 – 52.000	26dBW (50 – 51MHz, primary), 20dBW (51 – 52MHz secondary) (less for Foundation and Intermediate licensees).
USA	50.000 – 54.000	Band plan mandated by licence, 1500 watts.

Table 1.1: Examples of varying licence conditions for 6m operation.

tom end of the band for CW activity). Then CW up to 50.100MHz and CW / SSB above that frequency. By convention, 50.110 is the acknowledged inter-region calling frequency. This is helpful when not much is happening on the band, as your receiver can be left on this frequency which should be the first port of call of anyone seeking contacts when there is the possibility of a band-opening. The idea is that it is a calling frequency only and that once contact is established the stations should move to another part of the band. In practice this doesn't always happen and the frequency can quickly become chaotic. I have often heard stations in two or even three continents calling CQ or working stations simultaneously on 50.110, seemingly oblivious to one another and to the general confusion they are causing. Frankly, with *Cluster* available in most DXers' shacks and with many 6m DXers using transceivers with panoramic displays, it makes little sense to call on 50.110 once the band has opened. Call either side of that frequency (the 'DX Window' is considered to be 50.100 to 50.130MHz) and it won't be long before you are spotted by one means or another.

Higher in the band are designated meeting points for cross-band, meteor scatter, digital modes, SSTV, and other specialist interests. After that come designated channels for FM working, both simplex and repeaters, and priority channels for emergency communications. With 2MHz to play with in the UK (up to 4MHz in some countries, including the USA) there should be room for everyone, though the DX Window can fill quickly when there is a good band-opening. And it is important to be aware that some countries do not have the full band, while others are still limited to spot frequencies. **Table 1.1** shows 6m allocations in a number of countries to give a flavour of how they vary. It is not intended to be comprehensive. This list is maintained by Trevor Day, G3ZYY, and can be accessed on the Internet [3]. Some of the information may now have been superseded but the table shows clearly the way in which the various countries have taken different approaches to making 6m available.

EARLY 6M EQUIPMENT

Emil Pocock, W3EP, contributed a fascinating article on the history of 6m equip-

1965 ad for the Hallicrafters SR-46 6m transceiver.

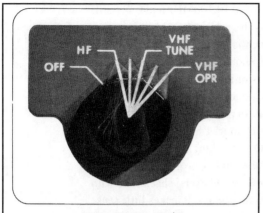

1965 ad for a Collins 'VHF Converter' to add 6m and 2m capability to the popular KWM-2 HF transceiver.

A 1kW homebrew amplifier for 6m, featured in the 1965 edition of *The Radio Amateur's VHF Manual*.

ment to *Six News* [4], and the following draws heavily on that material. One of the earliest sets designed specifically for 6m was the Gonset Communicator or 'Gooney Box' as it was affectionately known. This set was launched in the 50s and used a dual-conversion superhet receiver, and crystal-controlled transmitter producing about 10 watts. The price in 1955 was $230. The Heathkit Sixer was also popular during the 60s, costing just $45 in kit form. It provided 5 watts of transmit power (AM and CW) but the receiver was a regenerative design, sensitive enough but not exactly suitable for DX working. There were quite a few other models from manufacturers such as Lafayette, Hallicrafters, Knight and Clegg. Not surprisingly these were all US or Japanese companies, those being the two major countries at that time to have access to Six. The more serious 6m enthusiasts tended to build or buy a receive converter to work with their HF receiver and, for transmit, to build a separate, stand-alone transmitter unit, often crystal-controlled (those were the days when VHF operators would have just one or two crystals for transmit and, after calling CQ, would tune the band for possible replies).

Things improved enormously during the 60s when several US companies starting manufacturing VFO-controlled SSB rigs with as much as 100 watts output. These included the Heath SB-110, the Swan 250 and, for those wanting the Rolls-Royce of 6m rigs, the Drake TR-6. This boasted 19 valves, 10 transistors and 12 diodes and used three 6JB6 valves in the final to put out a hefty 300 watts. But the price was $600 in 1969, not cheap.

As many readers will recall, the face of amateur radio equipment changed dramatically in the 70s with the first solid-state equipment, coming mainly from Japan, the US manufacturers being somewhat slower in adopting the new technology. In 1976 Yaesu introduced the FT-620B, the first mass-produced, solid-state 6m transceiver.

The TS-600 and IC-551 were equivalent models, launched by Kenwood and Icom respectively. One of the great benefits of solid-state equipment was that it made portable operation so much easier, with equipment easily powered from a car battery rather than requiring a generator. Another way on to 6m was with a transverter. Yaesu, for example, sold a transverter frame to work with the FT-101 / 901 series of HF transceivers. Power was drawn from the transceiver and the frame could hold one or more VHF / UHF transverters, one of which could be for 6m.

By the 1980s 6m-only equipment was in decline and it was becoming much more common to find 6m as a featured band in both multiband HF and VHF transceivers. But perhaps the greatest boon to 6m operators was when manufacturers started to include general-coverage receive capabilities, which allowed DXers to follow the MUF upwards, perhaps by monitoring commercial and TV signals in the 45 – 50MHz range. Nowadays almost all new transceivers include 6m as standard.

THE 70MHZ – 4M – BAND

As mentioned previously, UK amateurs gained an allocation from 70.200 to 70.400MHz in 1956. The 4m Website [5] shows a QSO between G5KW and G8KW on 2 November that year as being a "first" for the band. Several other countries appear to have had access to the band during that period, with early "firsts" being shown for such countries as Algeria, Andorra and Eire. When this author gained his licence in 1968 the band was 70.1 to 70.7MHz. Nowadays it is 70.0 to 70.5MHz. For much of the 1970s, 80s and 90s, the only countries outside the UK to have access were countries such as Gibraltar and Cyprus plus, nearer to home, Eire and, farther away, South Africa. But all that has changed in the last few years, with a number of countries, mainly in Europe, the Middle East and parts of Africa, being granted 70MHz allocations. These haven't always overlapped, which has presented some challenges for making QSOs, but frequency allocations are gradually being harmonised.

SOME HISTORY

In the early days, 4m was very much a band for local contacts on AM and CW. Equipment was either homebrew or converted military surplus (which, in time, gave way to converted PMR equipment, still a popular way of getting on to the band with AM or FM). Activity within the UK was confined (largely because there wasn't the DX to be worked) to chasing squares and counties and the occasional contest. 4m activity in VHF Field Day in July was always of particular interest, with the hope that a ZB2 contact might be possible via Sporadic E. One of the first to

A rather untidy G3XTT shack from the early 1970s: an FT-101 with Europa transverter for 4m, plus an AR77 receiver and 2m Belcom Liner 2 SSB transceiver.

The Tom Withers TW4 transmitter.

offer commercial equipment for the band was Tom Withers with his TW4 transmitter, a 4m Nuvistor receive converter and 4m version of the popular TW Communicator transceiver. SSB activity followed on from the availability of SSB equipment for HF – my own early forays on to the band were with a Europa transverter (6146 PA) which was designed to interface directly to the FT-101 transceiver (which, fortunately, was what I was using at the time) and from which it derived its HT and other supplies. The late Jack Hum, G5UM, used his 'VHF' column in the *RSGB Bulletin* to encourage 4m activity although the band was never as widely used as might have been hoped (perhaps because, in many parts of the UK, there was still a clash with local TV frequencies).

WHAT DOES FOUR HAVE TO OFFER?

On a day-to-day basis there are many FM nets on 4m, usually run by local radio clubs. My own local club, for example, has been able to source and convert Ascom and, later, Simoco PMR radios for use on 4m, which has led to an active weekly net taking place among club members. But where things are changing almost by the day is the level of DX activity, as more and more countries gain access to the band.

Propagation on 4m is in many ways similar to 6m, with the exception of F2 propagation, which is rare to non-existent and, to date, no reported EME activity. (However, it must surely only be a matter of time before some enthusiasts start to make EME contacts on the band, especially given that antenna size for a given gain is more manageable than on 6m.)

The various other propagation modes to be found on 6m are in evidence on 4m, Sporadic E being responsible for much of the activity to date. In the past the Sporadic E season offered little of interest to UK amateurs active on 4m. If anything, the opposite was the case, Sporadic E bringing with it strong carriers from TV and other commercial stations from around Europe. Nowadays much of that has gone and, instead, there is the opportunity to work DX, mostly out to one hop (meaning that signals are good and simple equipment will more than do the job).

Stations in Southern Europe (for example) can take advantage of Trans-Equatorial Propagation to work over considerable distances on the band (the current distance record on the 4m website is EA6SX to ZS6WAB, a path of 7543km).

Meteor scatter is obviously available throughout the year when other propagation modes are not in evidence.

Over time we might expect more countries outside Europe to gain 4m allocations, at which time the DX chase will no doubt become more serious and very much more akin to what happens already on 6m.

In recent years there have been a number of portable operations activating rare squares on the band as well as activations by rare DXCC entities. For example, the Sovereign Military Order of Malta was active on 4m as 1A0KM in July 2007 during the Sporadic E season and again as 1A4A in October that year. It is also worth noting that although US stations are not licensed for 4m, a beacon operates each summer for propagation research. During 2013, for example, the beacon callsign (operating from Virginia) is WG2XPN and as usual it is on 70.005MHz from locator square FM07fm. Transatlantic propagation has occasionally been observed on 4m (see sidebar on page 29), so it might be hoped that US and Canadian amateurs gain access to the band before too long

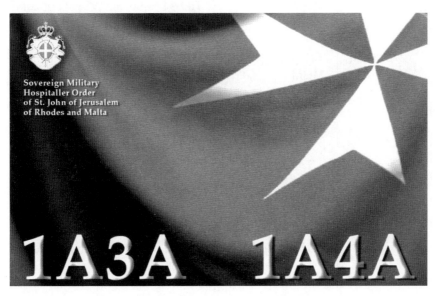

Sovereign Military
Hospitaller Order
of St. John of Jerusalem
of Rhodes and Malta

1A3A 1A4A

The rare DXCC entity of 'SMOM' (the Sovereign Military Order of Malta) was active on 4m in 2007.

A QUICK TOUR OF 4M

Frequency allocations and band planning on 4m are still evolving as new countries appear on the band, and it is worthwhile referring to the 4m website for the latest situation.

But, for a broad overview, the current IARU band plan for 4m gives a good idea of what to expect where – see **Table 1.2**. As can be seen, the pattern is very much the same as on 6m, with beacons and CW at the bottom end of the band, then SSB and a calling channel, with AM / FM and digital modes at the top end of the frequency allocation.

The cross-band calling frequency is worthy of particular mention. When few European countries had 6m access, cross-band working was a regular occurrence (usually 6m to 10m), but has largely disappeared. The same was and, to an extent, still is the case with 4m as amateurs with 4m access want to be able to conduct propagation experiments and make QSOs with countries that do not have that access. Again, this activity can be expected to reduce as more countries gain 4m access themselves.

FREQ	BANDWIDTH	MODES	USAGE
70.000 – 70.090	1000Hz	Telegraphy, MGM	Coordinated beacons
70.090 – 70.100	1000Hz	Telegraphy, MGM	Personal beacons 70.091 +/-500Hz WSPR beacons
70.100 – 70.250	2700Hz	Telegraphy, SSB, MGM	70.185 Cross band calling 70.200 CW / SSB calling 70.250 MS calling
70.250 – 70.300	12kHz	All modes	70.260 AM / FM calling 70.270 MGM activity centre
70.300	12kHz	All modes	70.300 RTTY / FAX
70.300 – 70.500	12kHz	FM channels, 12.5kHz spacing	70.3125 Digital communications 70.325 Digital communications 70.450 FM calling 70.4875 Digital communications

Table 1.2: The current IARU 4m band plan.

MAGIC BAND OPENS WIDE TO ALLOW CROSS-BAND TRANSATLANTIC CONTACT

(from the weekly ARRL Letter, 29 June 2007)

On Monday 25 June [2007] there was a big opening to Europe on 6 metres (50MHz), also known as the 'magic band'. In most areas, the opening started in the local US mid-morning, lasting until dark. According to *QST* column author of 'The World Above 50MHz', Gene Zimmerman, W3ZZ, "Many areas of the country that do not normally work Europe, including the Midwestern states, west Texas and Colorado, worked stations in Europe. At one point near the end of the opening, stations on the East Coast of the US were working stations in Hawaii on the Big Island." In this opening, Mike Smith, VE9AA, in New Brunswick was on 50MHz when he worked Nigel Coleman, G7CNF, on CW cross-band; Coleman was on the 70MHz band (4 metres) in England. Zimmerman said, "Though a few cross-band contacts were made via F2 propagation during the sunspot maximum period in the 1970s, this is believed to be the first 50 / 70 trans-Atlantic cross-band contact ever made on multi-hop Sporadic E propagation." RSGB VHF / UHF Manager David Butler, G4ASR, concurred: "It certainly isn't the first United Kingdom - North America 4 metre / 6 metre cross-band QSO, but it probably is the first via multi-hop Sporadic E." In North America, 70MHz is channel 4 on television sets.

SOURCES OF INFORMATION

This chapter has given a brief overview of the bands and covered some history. The rest of the book covers a wide range of topics from setting up your station for 4m and 6m, to operating techniques, contests and awards chasing and, perhaps most important of all, 4m and 6m propagation.

If you want to read beyond what is contained in these pages, there are many suitable sources. In terms of traditional publications, a number of excellent reference works are listed in the bibliography. Throughout the book there are also references to Internet sources. References pertaining to each chapter appear at the end of the chapter but **Appendix B** lists an extensive selection of sources. Internet URLs are valid at the time of going to press, though it must be remembered that Internet sites are appearing and disappearing all the time, while some change their locations. But surfing the net will take you to a huge number of useful sources; it would be quite impossible to list them all here. What this book will hopefully do is give you enough information to get you started and whet your appetite about aspects of Four and Six – whether it's propagation or antennas, history or DXpeditioning –that take your fancy and which you will want to explore further.

If you become a serious addict of Six you will almost certainly also want to join one of the 6m clubs such as the UK Six Metre Group. UKSMG publish an excellent magazine, *Six News*, run a very good website with lots of material available exclusively to members and organise social events where you can meet other enthusiasts and hear talks on DXpeditions and the like. They also sell a CD containing all back issues of *Six News*. This is well worth a read, perhaps during those times when Six is quiet!

There is no equivalent 4m club in the UK, though there is an Internet reflector for 4m enthusiasts [6].

Elsewhere in the world 6m enthusiasts join together, for example at the regular W6JKV gatherings, to swap war stories and hear advice from some of the experts. It is worth tapping into these networks if you become serious about Six. Not only will they give you access to expert advice but they are also a lot of fun. There are far fewer equivalent bodies or resources relating to 4m for the time being, but expect that to change as more people begin using the band.

REMEMBER THE VHF DXER'S OPERATING ETHIC:

1. If all the DX is S9, you're missing the *real* DX.

2. If the band is 'closed', find a way to open it!

(source: GM3SEK)

REFERENCES

[1] ZS6EZ website: http://zs6ez.za.org/lists

[2] UK Six Metre Group: uksmg.org

[3] List of 6m allocations: www.secornwall.pwp.blueyonder.co.uk/licensing.pdf

[4] Six Metre Transceivers, *Six News*, Issue 65, May 2000.

[5] Four metres website: www.70mhz.org

[6] UK 4m reflector: fourmetres@yahoogroups.com

2 Equipment and station design

I T MAY WELL be nowadays that you already have the necessary equipment to operate on Six. Almost every current transceiver includes 6m as a standard feature. This makes equipment selection much easier than in the past, when you generally needed to buy a dedicated 6m radio or a transverter which would take the output of your main transceiver, usually on 10m, and frequency-shift it to 6m. The same is not the case on 4m, though things are changing with at least two recently launched transceivers having 4m capability.

6M EQUIPMENT

The choice of equipment very much depends on what sort of 6m operating you have in mind. If all you want to do is some ragchewing with local amateurs, or you are looking to operate mobile, maybe working through a local repeater, then the main criteria may be minimum cost and small size. Neither of these should be a barrier nowadays. There are many excellent and small transceivers which cover a wide range of frequencies. Their new prices are modest considering the features you get for your money, and many perfectly adequate radios can be bought second-hand for a very reasonable outlay. You can buy a set specifically designed for channelised FM operation, such as the Yaesu FT-8900R, which is a quad-band (10, 6, 2m, 70cm) FM transceiver designed very much for mobile operation. Or you can have a multimode transceiver, most of which cover the HF bands as well as some of the VHF bands, good examples being the Yaesu FT-450D, the Icom IC-7000 and the Kenwood TS-480 which all cover HF and 6m with decent performance while being small and light enough for mobile or portable operation.

If, on the other hand, you are looking for a high-performance solution to 6m operation for DXing and / or contesting, then you really need to take a different approach. Using a transverter is still a viable way to get on Six if you already have a good quality HF transceiver. Ten-Tec of the USA used to sell a 6m transverter kit to work with a 20m transceiver, for example, while Yaesu sold the FTV-1000, designed to work with the FT-1000MP MkV transceiver but capable of operation with other transceivers, albeit requiring its own power supply when not mated with the FT-1000MP. Elecraft of the USA also sell a 6m transverter. Some transceivers are more suitable than others for use with an external transverter – if you

The Icom IC-9100 covers all bands from 160m to 23cm, with the exception of 4m.

are considering this route to getting on Six then ensure that your transceiver has a low-level transverter output.

In the past many multimode VHF / UHF base station transceivers such as the Yaesu FT-736 also offered excellent 6m performance. Such radios no longer appear to be manufactured, the line between HF and VHF equipment now tending to be drawn above 50MHz. So, for example, the Icom-910H offered 2m and 70cm base station capability (with 23cm as an option) although, interestingly, its replacement, the IC-9100, covers everything from 160m through to 23cm (except 4m).

On 6m, as on other VHF bands, a panoramic display is beneficial as it can be a handy way of alerting you to band openings. The Icom IC-756 series of transceivers has always been popular with 6m operators, partly for this reason and partly because they were one of the first mainstream transceivers to include 6m along with the HF bands. The IC-7600, which supersedes the 756 series, has the same capability.

But perhaps the biggest change in this respect in the past few years is that SDR (Software Defined Radio) equipment such as that from FlexRadio Systems now tends to have coverage extending to 6m and above. And many transceivers have an output in front of the main roofing filter, allowing connection to an external panoramic display such as the LP-Pan from TelePost Inc or the Elecraft P3.

What other considerations apply? If you are intent on DXing, then you will really need to be able to operate CW (brush up your CW if it's not a mode you currently use), and you will want CW filters in your transceiver where these are an after-purchase option. A general-coverage receive capability can also be useful so that you can listen outside the band for commercial signals which may give an indication of potential band openings. Most transceivers nowadays have this capability but SDR transceivers can be especially useful as you can literally 'watch' several pieces of spectrum below and up to 50MHz and, by doing so, see signals start to appear as the MUF increases (for example, during the onset of a Sporadic

E opening). A dual-watch receive capability is also useful for monitoring both sides of a DX pile-up.

On the transmit side, the question of power is an obvious one. Most of the base station and even mobile radios will now generate 50 or 100 watts on 50MHz which is more than adequate to get started on the band and to exploit many of the Sporadic E openings, during which signal strengths can be very high. But for long-haul and / or weak-signal operation it makes sense to be able to run as much power as your licence allows. This varies from country to country. In the US, for example, the power limit is 1.5kW whereas in the UK it is 400 watts at the antenna. This means that you are allowed to take feeder losses into account when considering what power you can run out of the transmitter. This said, it doesn't make sense, as will be discussed later, to accept high feeder loss and make up for it with additional transmitted power; this won't help on receive where the feeder loss will attenuate those weak incoming signals. But this advice from Ian White, GM3SEK, is equally pertinent, "*A bit more ERP will boost the station's capabilities into a higher league, where long-distance 'scatter' QSOs can be made almost any time, especially when the band is nearly open but not quite. If signals fade, just a one-way trace of signal from the station with higher ERP is often enough to keep the QSO alive until it can be completed. In difficult conditions, this can increase the completion rate quite dramatically. Given that most 6m users in the UK would not be able to increase their antenna size significantly, and the receive side is limited by external noise, then the power amplifier is the only way forward. Given also that most rigs can already deliver 100W, amplifiers of less than the 400 - 500W class are largely a waste of time and money*".

There is no single piece of advice which applies to the choice of a radio, as your selection will very much be determined by what you want. Do you operate portable and / or mobile? Do you want to be able to operate on other bands or only on 6m? Is it important to you to be able to monitor commercial frequencies outside the amateur bands in order to spot likely band openings? What modes are likely to be of most interest to you? And so on, including perhaps the most limiting criterion of all - what is your budget? The best way is to make a checklist of criteria and match it against the various radios on the market. And, where possible, track down the reviews that have been published by reputable bodies such as the ARRL and the RSGB. These will include laboratory measurements of performance and you will need to check out the 50MHz parameters.

The two tables, as compiled by *RadCom* reviewer Peter Hart, G3SJX, summarise a number of features of current transceivers that are relevant to 6m operators. Much of this data was acquired by Peter himself in the course of compiling his excellent reviews. **Table 2.1** shows data for multimode transceivers and **Table 2.2**, for completeness, some summary information about FM mobile and hand portable transceivers with 6m capability.

As a final comment on transceivers, it is worth noting the growing trend, already alluded to, to use an SDR (Software Defined Radio) receiver in conjunction with your main transceiver (or to use an SDR transceiver). There are several

Maker	Model	Type	Bands	VHF Rx coverage	Introduced	6m Tx Pwr	Supply	RadCom review	Other relevant 6m features
Icom	IC-706	mobile	HF 50 144	30-200MHz	1995	100	13V	November 1995	
Icom	IC-756	base large	HF 50	30-60MHz	1997	100	13V	May 1997	dual watch. Spectrum scan
Yaesu	FT-920	base large	HF 50	48-56MHz	1997	100	13V	August 1997	spectrum scan
Icom	IC-746	base mid	HF 50 144	30-60MHz	1997	100	13V	March 1997	
Icom	IC-706 mk2	mobile	HF 50 144	30-200MHz	1997	100	13V	June 1997	
Yaesu	FT-847	base mid	HF 50 144 430	30-76MHz	1998	100	13V	August 1998	dual watch. Spectrum display
Icom	IC-756PRO	base large	HF 50	30-60MHz	1999	100	13V	March 2000	
Alinco	DX-70TH	mobile	HF 50	50-54MHz	1999	100	13V	August 1999	
Icom	IC-706 mk2G	mobile	HF 50 144 430	30-200MHz	1999	100	13V		
Yaesu	FT-100	mobile	HF 50 144 430	30-970MHz	1999	100	13V	June 1999	spectrum scan
Kenwood	TS-2000	base mid	HF 50 144 430 1296	30-60MHz	2000	100	13V	April 2001	Data TNC.
Yaesu	FT-817	portable	HF 50 144 430	30-56MHz+	2001	5	9-13V	June 2001	int batteries/ext supply
Icom	IC-756PRO2	base large	HF 50	30-60MHz	2002	100	13V	June 2002	dual watch. Spectrum display
Icom	IC-7400	base mid	HF 50 144	30-60MHz	2002	100	13V	October 2002	spectrum scan
Yaesu	FT-897	base small	HF 50 144 430	30-56MHz+	2003	100	mains/13V	April 2003	20W with int batteries. Spectrum scan
Icom	IC-703	mobile	HF 50	30-60MHz	2003	10	9-15V	October 2003	transportable
Kenwood	TS-480HX	mobile	HF 50	30-60MHz	2003	200	13V	March 2004	Remote front panel
Kenwood	TS-480SAT	mobile	HF 50	30-60MHz	2003	100	13V		Remote front panel
Yaesu	FT-857	mobile	HF 50 144 430	30-56MHz+	2003	100	13V	June 2003	spectrum scan
Icom	IC-7800	base large	HF 50	30-60MHz	2004	200	mains	August 2004	twin receivers. Spectrum display
Yaesu	FT-2000D	base large	HF 50	48-54MHz	2004	100	13V	February 2005	twin receivers. Spectrum display with DMU
Icom	IC-756PRO3	base large	HF 50	30-60MHz	2005	100	13V	December 2006	twin receivers. Spectrum display
Yaesu	FTDX9000	base large	HF 50	30-60MHz	2005	200	mains	December 2006	twin receivers. Spectrum display
Icom	IC-7000	mobile	HF 50 144 430	30-200MHz	2005	100	13V	April 2006	spectrum scan
FlexRadio	SDR1000	SDR	HF 50	30-65MHz	2006	1	13V	June 2006	External PC and soundcard needed
Icom	IC-7700	base large	HF 50	30-60MHz	2008	200	mains	July 2008	twin receivers. Spectrum display with DMU
Yaesu	FT-2000	base large	HF 50	30-60MHz	2007	100	13V	March 2008	twin receivers. Spectrum display with DMU
Elecraft	K3	base small	HF 50	48-54MHz	2008	100	13V	August 2008	twin receivers. Spectrum display
FlexRadio	5000A	SDR	HF 50	30-65MHz	2008	100	13V	September 2007	Remote operation via LAN. Spectrum scan
Yaesu	FT-950	base small	HF 50	30-56MHz	2007	100	13V	December 2007	Spectrum display with DMU accessory
Yaesu	FT-450	base small	HF 50	30-56MHz	2007	100	13V	January 2008	
Ten-Tec	OMNI VII	base mid	HF 50	48-54MHz	2007	100	13V		External PC with fireware interface needed
Ten-Tec	Eagle 599	base small	HF 50	50-54MHz	2011	100	13V		External PC with fireware interface needed
Icom	IC-7200	base small	HF 50	30-60MHz	2008	100	13V	January 2009	Weather resistant and rugged
FlexRadio	3000	SDR	HF 50	30-60MHz	2009	100	13V	June 2009	twin receiver option
Icom	IC-7600	base large	HF 50	30-60MHz	2009	100	13V	August 2009	Spectrum display with DMU
Kenwood	TS-590S	base mid	HF 50	30-60MHz	2010	100	13V	January 2011	Spectrum scan
FlexRadio	1500	SDR	HF 50	30-54MHz	2010	5	13V	April 2011	External PC with USB interface needed
FlexRadio	3000	SDR	HF 50	30-60MHz+	2010	100	mains	June 2010	twin receivers
Icom	IC-9100	base mid	HF 50 144 430 1296	30-54MHz	2011	100	13V	May 2012	satellite and crossband duplex, DSTAR
Yaesu	FT-450D	base small	HF 50	30-56MHz	2011	100	13V	November 2011	
Icom	IC-7410	base small	HF 50	40-56MHz	2011	100	13V	January 2012	spectrum scan, RTTY decoder
Elecraft	KX3	portable	HF 50	30-77MHz	2012	10	8-15V	April 2013	int batteries/ext supply
FlexRadio	6000 series	SDR	HF 50	44-54MHz	2012	100	mains		Up to 8 receivers
Kenwood	TS-990S	base large	HF 50	30-60MHz	2013	200	mains		twin receivers. Spectrum display
Yaesu	FTDX3000	base mid	HF 50	30-56MHz	2013	100	13V	June 2013?	Spectrum display
Icom	IC-7100	base small	HF 50 70 144 430	30-200MHz	2013	100	13V		Remote front panel

Table 2.1: 6m multimode transceivers covering SSB / CW and in most cases AM and FM.

Maker	Model	Type	Bands (MHz)	Year Introduced	6m TX power (W)	*RadCom* Review
Alinco	DR-M06	Mobile	50	1995	20	April 1995
Yaesu	FT-8900R	Mobile	28 50 144 430	2002	50	
Icom	IC-E90	Handheld	50 144 430	2004	5	December 2004
Icom	IC-T81E	Handheld	50 144 432 1296	2000	5	September 2000
Icom	IC-T8E	Handheld	50 144 432	1997	5	
Yaesu	VX-5R	Handheld	50 144 432	2000	5	September 2000
Yaesu	VX-7R	Handheld	50 144 432	2003	5	October 2003
AKD	AKD-6001	Mobile	50	2000	25	
Yaesu	VX-8DE	Handheld	50 144 432	2008	5	

Table 2.2: 6m FM transceivers.

Some of the new range of HF + 6m transceivers that have come on the market in the last few years (see Table 2.1 opposite). Top: Kenwood TS-590S; Middle: Kenwood TS-990S; Bottom: Yaesu FTDX5000. These three transceivers are here shown to scale.

Photo: funcubedongle.com

The FUNCube Dongle is an SDR in USB form, requiring no drivers.

benefits. The ideal situation is one on which the SDR is able to access the whole, or a significant part, of the band, which can then be presented in graphical format on your PC screen via software such as *WinSDR* [1]. This allows a bandscope display of the frequency segment being monitored, so that you can see immediately if signals are being received. Use it also with software such as *CW Skimmer* [2] and you will even see the callsigns of stations being copied. Before this book goes out of print, it is by no means beyond the bounds of possibility that an SSB version of *CW Skimmer* will appear! If the SDR receiver is fed after the roofing filter in your transceiver much of the benefit is lost, as you will then only be able to display and monitor 3kHz of the band. As a 6m enthusiast you really want a capability not only to monitor quite a lot of spectrum, but for this to go well below 50MHz, so that you can be alerted to potential band openings.

An interesting postscript to the topic of SDR is the recent availability of cheap SDR devices in USB form. These range from TV tuners sold for just a few pounds on eBay and elsewhere (The Realtek RTL2832U Radio & TV Dongle, for example) to products like the FUNCube Dongle [3], designed by Howard Long, G6LVB. The former requires some additional software to be loaded to be usable as a broadband SDR receiver (just do an Internet search for details); the latter requires no additional drivers at all. Neither device is comparable to a full-spec SDR receiver or transceiver, but both offer the opportunity to monitor a wide range of frequencies easily and conveniently for relatively low cost.

4M EQUIPMENT

For many years the route UK amateurs took to become active on 4m was to convert PMR equipment (for AM or FM) or, in the case of SSB and CW, to build, usually a transverter rather than a 4m transmitter as such (although it was not uncommon to have a receive converter linked to, usually, a 10m receiver while, at the same time, using a separate transmitter specifically designed for Four). As was mentioned earlier in this book, there were one or two pieces of commercial equipment designed specifically for the band, but the selection was very limited.

Until recently, no multiband transceiver has included a true 4m capability, although the Yaesu FT-847, as sold in the UK, had a limited capability on Four. That is to say, it can run about 20 watts out but with poor efficiency and is generally considered somewhat 'deaf' on receive. A number of modifications exist on the Internet to improve both receive sensitivity and transmit efficiency and power output and at least one of the major UK distributors of Yaesu equipment

offers to install their own version of the transmit modi-
fication for a modest charge, thereby achieving
higher transmit power output for less internal heat
dissipation.

Nowadays a few transverters are still manufactured, invariably solid state and generally running at about the 25 watt level. One example is the
Spectrum Communications product, which is available in versions with a 2m or 10m IF. With the release of
the band to other European countries, similar products are now being manufactured elsewhere,
transverters from Italy (IK1ZYW) and Hungary
(HA1YA) for example, and amplifiers by Amplitec
[4] in Hungary and others. If you are prepared to
build your own, a number of designs appear, for example in the *RSGB VHF / UHF
Handbook* [5] and OZ2M [6] has published a design which is available in kit form.

The new Icom IC-7100 includes 4m as
standard: no doubt other manufacturers
will follow suit.

Perhaps the most interesting development in the past year, though, has been
the announcement of the Icom IC-7100 transceiver which, at least in its European
variant, will include 70MHz as standard. This news will almost certainly be
followed by the other major manufacturers in due course, or so we can hope.
Indeed, just prior to going to press, there was news of the launch of a dual-band
50 / 70MHz SSB / CW 20 watt transceiver (the 6N4) from Noble Radio, a start-up
venture from the USA.

The latest FlexRadio 6000 series of equipment also has coverage extending
above 70MHz and offers some extremely interesting possibilities, with the ability
to monitor multiple blocks of spectrum simultaneously. The downside is that the
price reflects the very high specification.

**Looking as smart as a piece of high-end hi-fi equipment, the new Flex-6000
series of SDR transceivers have coverage extending beyond 4m. Is this the
future look of all amateur radio equipment?**

LINEAR AMPLIFIERS

As with transceivers, nowadays there are many linear amplifiers which operate on 6m. Some are single-band units, designed specifically for Six. US companies such as Henry have manufactured these in the past and Alpha is currently offering the 8406 model. M2 manufactures a solid state unit which could prove to be ideal for portable operation. The Lunar-Link Systems LA-62A by K1FO (SK) and the Commander 1200 from Command Technologies are two other popular choices.

Others are multiband, usually HF amplifiers with a 6m capability. There are many examples, both valve amplifiers such as the Acom 1000, and semiconductor units such as the Yaesu VL-1000 Quadra, SPE Expert 1K-FA and 2K-FA and the Tokyo Hy-Power HL-1.5Kfx.

The OM Power OM1500, manufactured in the Slovak Republic, is one of several new linear amplifiers which cover the 50MHz band.

New models covering 6m are coming on the market all the time as the popularity of 6m continues to grow. OM Power has recently introduced several covering 6m, including the OM1006 transistorised amplifier and the OM2006 and OM3006 valve amplifiers, the latter reviewed by G4BWP / A65BD and G8VR in *Six News*, Issues 110 and 111. Their new OM1500 covers all the HF bands as well as Six and, at 22kg, is described as "the smallest and lightest 1500W PA on the market". Acom too released the Acom 1500 in 2012. It also covers HF + 6m and was reviewed by Peter Hart, G3SJX, in the December 2012 *RadCom*.

Any of these amplifiers will be perfectly suitable for 6m operation, the main trick with amplifiers being to avoid overdriving them, so it always pays to monitor your signal or have a local amateur check it for you (but if you are very loud with him, it is important that he can differentiate between a poor signal on your side and simple overloading of the receiver on his side). Generally speaking, it is worth buying an amplifier capable of producing more power than you will actually want to run, so that you can always operate it well within its specification, for cleanest signal and longest life. And if you intend to operate datamodes such as WSJT, the high duty cycle again means that the amplifier should be over-rated to avoid early failure.

Building your own amplifier is a major undertaking, but designs do appear in various places, including the RSGB *VHF / UHF Handbook* [7] and there is a very popular design for an expedition amplifier on the G3WOS web page [8].

Amplifiers for 4m also exist. There are some ex-PMR commercial models (the solid-state A200 being one such) which are easily converted to 4m and

Easily converted to 4m operation, the Pye A200 amplifier is often available on the surplus market.

provide useful power levels of tens of watts, but some manufacturers, in particular Linear Amp UK, offer combined 6m / 4m amplifiers capable of well over the UK power limit meaning that they can be run comfortably at the limit even when in high duty cycle (datamode) applications.

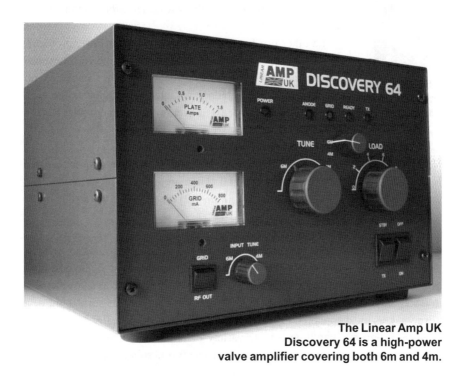

The Linear Amp UK Discovery 64 is a high-power valve amplifier covering both 6m and 4m.

ANCILLARY EQUIPMENT

In terms of other equipment, this will depend on circumstances. Strictly speaking, the UK licence still requires a means of checking fundamental frequency and harmonics independent of your main transceiver, usually some kind of wavemeter, though it is doubtful that many shacks have such equipment nowadays. But a second (general coverage) receiver to allow you to check for spurious emissions and generally to monitor your transmitted signal is always a good idea.

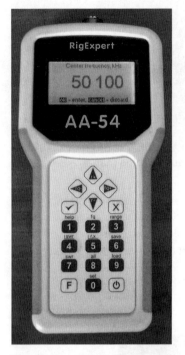

The RigExpert AA-54 antenna analyzer, which covers frequencies up to 54MHz.

Your transceiver may well have a built-in SWR meter, which will allow a basic check on whether the antenna is working correctly. Otherwise an external SWR / power meter is a valuable asset in the shack. If you are going to do anything more than put up a single Yagi you may also want some more specialist antenna test equipment, for example a swept analyzer such as the miniVNA or the RigExpert AA-54, both reviewed in *QST* [9, 10]. Such an analyzer will not only allow you to check antenna SWR and impedance, but also feedline loss, and will allow you to cut phasing lines, for example, when building a stacked array.

Other equipment that you might want to consider includes filters of various descriptions. In the shack you may wish to install a low pass filter on the output of your transmitter (or linear amplifier if the filter can handle the power) to reduce any harmonic radiation to a minimum (but do ensure it is designed for 50MHz or 70MHz operation). Designs appear in a number of places such as the *ARRL Handbook* and on the Internet [11].

The other use for filters is when you have EMC problems with a neighbour's electronic device, at which time you may need to install one or more varieties of band-stop filter, choke etc on the various ports of his equipment (antenna lead, power line, etc). This topic is further discussed later in this chapter under EMC.

COMPUTERS AND SOFTWARE

No shack is complete nowadays without a PC. The PC has many uses and this is not the place to describe them in detail. They are covered in some depth in the RSGB book, *Computers in Amateur Radio* [12] for example, but also in many other places.

Briefly, there are two areas of use, local and the wider network. Locally, you will use the PC for logging, for antenna modelling, for running software such as propagation forecasts (though this is not too useful at 50MHz) and, increasingly, as a bandscope in conjunction with an SDR receiver (or, indeed, as a key element in a station designed around an SDR transceiver). 'Logging' is something of a misnomer, as most so-called logging programs nowadays are really *station management* programs, handling a wide variety of tasks including such things as rotator control, *Cluster* interface, Logbook of The World submission, award tracking etc. 4m enthusiasts should be aware that many logging programs do not recognise the 4m band, so should bear this in mind when selecting logging software. The same applies to contest logging software. For RSGB contests, *SDV* (the VHF version of the EI5DI's *SD* software) does the job nicely, but is certainly not the only option available.

You will probably also want to use the PC for running datamodes software such as WSJT, described in the specialist modes operating section of this book.

What does all this mean for the PC's specification? For use for Internet access and basic logging functions a PC retired from other functions within the home may be perfectly adequate. For use as part of an SDR capability, a fast PC with high-specification soundcard will be required. The good news is that the price of such computing capability has fallen dramatically in recent years.

As far as the operating system is concerned, most software for amateur radio use is designed to work with the *Windows* operating system as this is by the far the most ubiquitous. But beware that, because much of this software has been designed by fellow amateurs, they may not always have time to upgrade it to work with the very latest version of *Windows*, so it is always worth checking. And some pieces of ancillary equipment, including transceivers themselves, still use a serial port rather than a USB connector, whereas very few modern PCs have a serial interface, so an adapter will be required. Some work better than others so beware which chipset and driver program your USB / serial adapter uses – there is a lot of useful information available on the Internet if you run into difficulties.

THE INTERNET AND TELEPHONE ALERTING

It may seem obvious, but nowadays you will want the Internet to be integrated into your station. There is so much that it puts at your fingertips, from the *DX Cluster* system to the ON4KST chat room, from real-time propagation data to world locator square maps, and much more. The use of these tools is described in Chapter 5. But simply having the Internet on a screen by your transceiver may not be enough. After all, do you spend all your time in the shack? It has been said many times in this book and elsewhere that 6m and 4m openings can be fleeting and, Murphy's Law being what it is, will invariably occur just as you have gone downstairs, into the garden or to the shops. So it is worth thinking about how to deal with these situations. In the past, 6m DXers would have shared lists of what they needed and, if a 'Wanted' one had appeared, would use the phone to alert each other to the DX. This may well still happen on a

G7VJR's *DXLite* is designed to be monitored on a Smartphone.

localised basis but there is the general assumption that everyone has the Internet nowadays and that other forms of alerting are unnecessary. What to do? A good start is to ensure that, when there is a rare one scheduled to be on, you have the Internet with you wherever you go. If you have a wi-fi network in the house this should be easy. Use a laptop and carry it from room to room. Your wi-fi network may even extend to the garden. Or for ease of use, carry a wi-fi capable device in your pocket (almost certainly a Smartphone of some sort) and connect via one of the fast Cluster-like services such as *DXLite* server developed and run by Michael Wells, G7VJR, designed with PDA and mobile phone users in mind [13], or with a an altering program such as *DX Hunter* [14] (designed for the iPhone /iPad / iPod operating system) which, when programmed against your needs list, will only alert you if a *Cluster* spot meets your pre-determined criteria.

The other way to be alerted while you are out and about is to have the Internet, working in conjunction with your logging software, alert you by way of a text (SMS) message to your mobile phone or a transmission to your VHF handheld. Many logging software packages have the capability of doing either or both of these things, to a set of predetermined criteria (e.g. a new one on the band, or a new mode slot on the band).

DEALING WITH EMC PROBLEMS

Although Band 1 TV is no longer an issue in most parts of the world, anyone planning to chase weak-signal DX on 6m or 4m will almost certainly be faced with EMC issues, particularly when running high power. The most common problem is with masthead and other distribution amplifiers for FM radio and TV. Many of these are broadband in their characteristics, exhibiting significant gain at 50MHz.

The solution has to lie at the neighbour's location when such problems occur. After output filtering, there is nothing that you can do in your own shack, other than reducing power, which is unacceptable if you are to be ready to chase DX when those short openings occur. The solution may require several elements and each case will be different. The sort of options to investigate include:

- Dealing with common-mode signals (induced directly into the feeder) at the input to the amplifier. This can be tackled by installing common-mode chokes in series with the input to the preamplifier. Such a choke can consist of 12 turns of miniature coax (e.g. Maplin XR88) wound on Fair-Rite 2643802702 cores, as described, for example, in the *RSGB Yearbook*.

- Dealing with differential-mode 50MHz signals (picked up by the TV antenna) at the input to the amplifier. One source recommends a lowish-Q parallel resonant circuit (500nH and 20pF) in-line between the common-mode choke and the input to the amplifier. This has over 30dB attenuation at 50 MHz but less than 3dB attenuation at 90MHz.

- Dealing with 50 or 70MHz signals reaching the amplifier via the DC voltage feed. The solution to this is a common-mode choke on the DC voltage line.

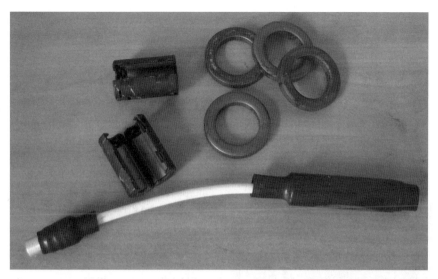

Weapons in the fight against EMC problems: clip-on ferrite cores, ferrite rings. and a commercially-made high-pass filter suitable for fitting to VHF / FM radios, TV receivers, distribution systems etc.

You may find your transmitted signals getting into other equipment, and similar advice applies, but in the UK masthead and distribution amplifiers are by far the most common problem. Unfortunately they require the neighbour to be co-operative, given that the solution has to be at his location. In those countries where cable TV is widespread, the major EMC problems with neighbours are likely to arise from other consumer devices such as stereo systems, computer speakers and the like. These are usually indicative of inferior design of the product and may not be simple to eliminate. In general, though, EMC problems *to* neighbours are much less of a problem than they were in years past.

You may also suffer incoming EMC problems if you are in a noisy location, preventing you from pulling through the weakest signals. Dealing with these requires a portable receiver and the patience to walk the neighbourhood to track down the offending device(s). The solution will depend entirely on what it is you find and, again, co-operation from the neighbour(s) concerned.

Remember, though, that a major contribution to reducing noise can be achieved simply through having a good antenna with narrow beamwidth, which will go a long way to reducing incoming noise (except, of course, when you have to beam in the direction of the noise source!) Stacking Yagis, as well as improving gain and take-off angle, is also a great way of improving your signal-to-noise ratio. Costas, SV1DH, who lives in a noisy urban environment, has experimented with an ANC-4 wide-band noise canceller (using a 5-element Yagi as the noise an-tenna) but not found this effective most of the time. Such a system can suppress the noise from a single source by up to 20dB, but often in such an environment one is faced with multiple noise sources. Some 6m DXers report good results with the DX Engineering NCC-1 Variable Phasing Controller. And one advantage of

Software Defined Radios is that, as noise-cancelling algorithms improve, the relevant software can readily be downloaded and incorporated. Already, for example, a *Windows* version of the SM5BSZ noise elimination system has been incorporated into the operating software of the SDR-1000

DEALING WITH NOISE SOURCES

LF operators are used to living with high levels of background noise, both natural and manmade. VHF operators expect to have low noise levels and to be able to deal with extremely weak signals. But, increasingly, man-made noise is becoming a problem in many locations. There are several ways of mitigating the effects, although very few operators will be able to escape the problem entirely.

The noise level on 6m at the G3WOS station can clearly be seen on the IC-7800 S-meter and spectrum scope.

Chris, G3WOS, tracking down local noise sources using a portable receiver and a small Yagi. He comments, "This is definitely best done in the dark..."

One solution, to which an increasing number of operators are turning, is to have a remote station in a rural (quiet) location, which they can run from home. Remote operation is covered elsewhere in this book, as it has other uses too (for example, being able to run your home station when out and about).

But what to do if you are faced with manmade interference at your main station and want to be able to work DX from there? Sometimes the interference is from specific directions which are of little or no importance. Life being what it is, though, it will almost certainly emanate from one or more key directions!

One 6m DXer to have made a specific study of the matter is Chris, G3WOS. Chris found that the increasing levels of noise around his home were playing havoc with his 6m DX activities, particularly EME where signals can be very weak indeed and, when trying to exploit ground gain, it is necessary to work at the lowest take-off angles (and therefore be beaming right through the surrounding urban area).

Although there has been a lot of talk about the problems of power line adapters (for datacomms distribution within houses over mains wiring), Chris found that the worst offenders were switched-mode power supplies, plasma TVs and low-voltage downlights. These various sources can often be tracked down using a portable receiver and simply walking the neighbourhood. The problem then becomes one of persuading the offending household(s) that there is a problem which needs to be attended to. This is where many amateurs find it hard to take the next step as it involves diplomacy rather than technology. It's really a matter of whether the problem is affecting your DXing enjoyment to the extent that drastic measures are called for.

In reality, if approached the right way, many people are prepared to listen and to cooperate, and the solutions are often straightforward and can be portrayed as helping to solve a potentially dangerous situation (for example, an enclosed power supply that may be overheating and therefore potentially a fire hazard).

Chris gives some valuable hints and tips (along with case studies) on his own website [15] and also refers to ON4WW's website [16] which, again, contains lots of useful information. And UK amateurs should also remember that the RSGB EMC Committee is there to help. There is often a Committee member living not too far away who can tap into a wealth of experience and advice. The moral of the story is to persevere in dealing with such issues, rather than allow them to adversely impact your enjoyment of the hobby.

REMOTE OPERATION

Given the difficulties in building a really competitive 6m station in many home locations (surrounding hills, lack of space, planning limitations, high local noise levels, EMC problems etc), a number of serious 6m DXers have looked for a workaround by setting up a station elsewhere. This may be at a friend's location or one owned by the amateur himself but remote from the main home. One well-known UK 6m DXer, for example, has an old World War II bunker where he has

The Ten-Tec Omni VII, which Ten-Tec describes as "the first truly Net-Ready ham transceiver". No PC is necessary at the transceiver in order to operate the station remotely. The transceiver can be located anywhere that has wideband Internet access.

The Flex-5000A SDR is literally a 'black box', with a blank front panel but for the headphone and microphone sockets. It has a built-in computer interface and networking capability, making it ideal for remote operation.

installed his main 6m station, not too far from his home but on a much more advantageous site. The benefits of such an approach are obvious if it results in a more effective station but there are disadvantages too, usually in terms of the time it takes to get to the site when a DX station appears. One way around this is to arrange to be able to operate the remote station from elsewhere, either from your home or from your work. The benefits also work in the reverse direction too. There can be nothing more irritating than being stuck at work or at a friend's or relative's house and knowing that you are missing a rare one on Six! But many 6m DXers nowadays get round this by operating their home station remotely, to avoid missing those fleeting openings.

Remote operation is much easier nowadays than in years gone by. The ubiquity of the Internet offers a ready-made networking solution, always on, and many transceivers nowadays come ready to interface to the Internet. Good examples are the Kenwood TS-480, the Elecraft K3, the Ten-Tec Omni-VII and, perhaps most obviously, various Software-Defined Radios such as the Flex-5000A which are not simply built with a networking capability but have the computer interface as an inherent aspect of their design and operation.

Remote station operation, though, requires more than being able to control the transceiver remotely. You need to be able to key the radio for CW or send and receive voice for SSB. You also need to be able to rotate the antenna and control the amplifier. Several articles have dealt with these aspects (e.g. [17]) and if you intend to go down this route you would be well advised to consult some of those sources and do some thorough planning based around your specific requirements. But don't be put off as remote operation is getting easier all the time, and there are now several vendors who supply the equipment you will need at either end of the link in order to control not just your transceiver but other items such as rotator and linear amplifier. There is lots of information available on the Internet and several books (for example [18]) covering this increasingly important and accessible topic.

A final word of warning on remote operation, though. Do ensure that you set up the capability in a way which is within the terms of your licence (licensing authorities tend to be nervous about remote operation because there is always the potential for a carrier to be transmitted under fault conditions with the remote operator unable to turn off the transmitter). Also, the rules and regulations pertaining to certain awards and contests rule out remote operation for those particular activities, so do please check that you fall within the relevant restrictions (for example, at the time of writing the DXCC rules prohibit contacts made where the remote operator is in a different DXCC entity to the station).

OTHER CONSIDERATIONS

There is much more that could be said about designing and building your station, but they are considerations that are by no means specific to 6m and 4m operation. Subjects such as station layout for operating convenience and comfort. Safety concerns, both electrical (for example, an easy-to-access main switch) and mechanical (particularly where antenna masts are concerned). Cable routing for minimal noise pickup. Good earthing. Station security and insurance. And so on. These are covered in other sources, for example in the *RSGB Operating Manual*. In the present context, a significant factor may be whether your station is engineered purely for 6m and / or 4m, or whether you operate other bands, too. If the latter, is your 6m and / or 4m ready to go, given that band openings can be fleeting and happen at short notice, or is a certain amount of station reconfiguration required?

The newest transceiver covering HF plus the 6m band (at the time of writing): the Yaesu FTᴅx1200. Sitting somewhere between the FT-950 and the FTᴅx3000 it terms of both features and cost, the FTᴅx1200 is to be released in mid-2013.

But perhaps the most pertinent piece of advice is to keep a station notebook and record details not only of station wiring and configuration (plus, perhaps, equipment serial numbers and the like) but also make a note of key operating parameters such as cable loss. This is especially important in weak-signal VHF work. If you are able to characterise your feeders and then re-measure them on, say, an annual basis, you will quickly see whether their performance is suffering and they need to be replaced. Moisture ingress in particular can be insidious and not visible to the eye, but can seriously affect cable loss. Connectors can also deteriorate over time, especially if the connections were not well made or sealed in the first place. For VHF use, it is never a good idea to buy poor quality connectors – money spent on good ones (probably the pressure-seal type) is money well spent. And any connections should also be well water-proofed. There is a lot of advice available on the best way to do this while ensuring that you can remove the waterproofing easily should the need arise. Various forms of self-amalgamating tape, wrapped over with regular insulating tape, are commonly used and can be removed with a sharp knife. The author has a different approach which is to wrap connectors with Blu-Tack™, a recommendation which was given to me some years ago by a radio amateur who lived close to the sea and had found that this was about the only sealant that seemed to be oblivious to the ravages of a salt-laden atmosphere. Similar products (the normal purpose is for attaching posters to walls and similar) are no doubt available in other parts of the world.

REFERENCES

[1] *WinSDR*: http://psn.quake.net/WinSDR

[2] *CW Skimmer*: www.dxatlas.com/CwSkimmer

[3] Fun Cube Dongle: www.funcubedongle.com

[4] Amplitec: www.amplitec.hu/ug_4_100_1000_gs31b_eng.html

[5] VHF / UHF Handbook, RSGB, p103.

[6] OZ2M transverter design: www.rudius.net/oz2m/70mhz/transverter.htm

[7] *VHF/UHF Handbook*, RSGB, p108.

[8] G3WOS 6m amplifier design: www.gare.co.uk/amplifier/index.htm

[9] 'Two More Antenna System Measurement Devices' *QST*, August 2008, pp43-47.

[10] 'A Look at Four Antenna Analyzers', QST, March 2012, pp46-52.

[11] Low Pass Filter: www.ham-radio.com/n6ca/50MHz/50appnotes/50tlpf.html

[12] *Computers in Amateur Radio* (2nd edition), edited by Steve White, G3ZVW, RSGB 2013.

[13] *DXLite*: http://g7vjr.org/dxlite

[14] *DX Hunter*: www.michiv.de/dxhunter

[15] '[Not] Living with 6m/50MHz interference!': www.gare.co.uk/noise/index.htm

[16] ON4WW EMI - RFI page: www.on4ww.be/emi-rfi.html

[17] 'There's a Remote Possibility…', David Gould, G3UEG, *RadCom* August 2005 p80, September 2005 p76, October 2005 p76.

[18] *Remote Operating for Amateur Radio*, Steve Ford, WB8IMY, ARRL 2010.

3 Antennas for 6m and 4m

YOUR INTEREST IN 6m or 4m may be to work local friends on FM, either simplex or via a repeater. For this you will usually use a vertical antenna, to give all-round coverage and because vertical polarization is used for this purpose (primarily because it is easier to install a vertical antenna for mobile operation and also because, for local nets, it is important to have all-round coverage). For base station use you will probably choose a quarter- or half-wave vertical antenna, mounted at rooftop height or maybe higher (for example, if you already have a tower supporting antennas for other bands). There are many suitable commercial antennas available for this purpose. Some are multiband, so that you only have to install a single antenna to cover several of your favourite bands.

For DX working, you will need a horizontally-polarized antenna with a useful amount of gain, though the strength of Sporadic E signals, as has been mentioned elsewhere, is often such that a very simple antenna, maybe just a dipole, will be more than adequate. But for chasing those weak signals, a Yagi antenna of probably five elements or more is essential, mounted at the greatest height you can achieve in order to have a very low take-off angle. If your previous experience has been on, say, 2m, then a 6m Yagi in particular will probably seem rather large, whereas if you are coming to 6m from HF it won't be so challenging either to you, the neighbours or your XYL!

On 4m, even a 7-element Yagi is quite a manageable size, and even for a single-operator portable station. This is OH0/OZ2M on 70MHz from the Åland Islands, May 2012.

Justin, G0KSC, of InnovAntennas giving a talk on antenna design at G3WOS's '6m BBQ' in 2011.

A huge amount has been written about the merits and demerits of different antenna designs for 6m, as well as about the use of a single Yagi compared with a pair of Yagis stacked. Your choice may well be governed by practical issues, such as what it is possible for you to put on an existing tower in addition to antennas already *in situ*. Suffice to say that a poll of serious 6m DXers would show that the majority seem to favour a single long-boom Yagi of, perhaps, six or seven elements, or a stacked pair of Yagis. Where circumstances allow, a stack is probably the ideal solution, as discussed later in this chapter. In practice, relatively few amateurs are in a position to build one because their tower is likely already to be carrying other antennas such as an HF tribander and / or Yagis for the higher VHF bands. For 4m, given that most activity is within one hop of Sporadic E, a single Yagi of 3 elements or more is probably more than adequate.

2 x 6-element 50MHz InnovAntennas LFA Yagis at the station of VK6OX in Western Australia.

Other designs of antenna can and do work on VHF, as you would expect – even wire antennas such as longwires or loops of various types. But, with the exception of a vertical antenna, in practice you will want to be able to rotate your antenna which implies a more rigid design than can be achieved with wires. Some band users have successfully built quad antennas using garden canes and wire, but in practice some sort of aluminium construction is likely to be the order of the day. There are many good designs which you can build yourself and many more which are available commercially. The first consideration, though, is what your aspirations (and budget!) might be.

One area in which there has been some interesting progress in recent years is Yagi design, with InnovAntennas [1] in particular developing some new and interesting designs. Justin Johnson, G0KSC, has developed two main types of new Yagi design, which he calls the LFA (Loop Fed Array) and the OWL (Optimised Wideband Low impedance).

The LFA Yagi has a rectangular shaped, full-wave loop driven element that is laid flat on the boom between and in-line with the parasitic elements. The smaller end sections of the loop driven element, which run parallel to the boom, are designed to be 180° out-of-phase with each other, therefore each side cancels the other out and so minimum radiation occurs from them. The main benefit is the rejection of unwanted noise, due to the highly suppressed side lobes combined with a particularly broadband design.

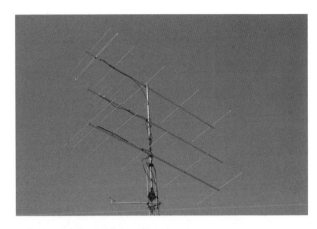

A stack of two 4m 6-element OWL Yagis (with a 2m LFA between).

G0KSC's Optimised Wideband Low impedance ('OWL') Yagi is a low-impedance (12.5Ω) design in which the split dipole driven element is changed to a folded dipole to transform the feed impedance to 50Ω. This allows for direct feed with 50Ω cable and removes any matching losses. While the use of a folded dipole in a Yagi is nothing new, it is generally done with 50Ω split-dipole designs to give a 200Ω feed impedance, which then requires a 4:1 balun or other matching device in order to be fed with 50Ω cable. G0KSC has computer optimised the whole antenna with its folded dipole to provide a design with no matching losses and high front-to-back ratio.

Up to a point, then, you can now select a Yagi to suit your particular situation, whether it's a location where you want to be able to null out particular local noise sources or a transmitter that doesn't like running into anything other than a very low SWR.

Of course, when choosing an antenna there are two factors to consider. One is the electrical performance, the other is physical design. Some expeditioners have gone to great lengths to develop Yagis which break down into short lengths and are (relatively) lightweight for their assembled size and performance. For home station use you would probably be much more concerned that the antenna be solid in its construction and able to withstand the worst your local weather can throw at it over a period of years.

WHAT AM I TRYING TO ACHIEVE?

There is an old adage in amateur radio which is that one can never have enough antennas. There is another, equally old, one which says that if your antenna didn't fall down last winter it clearly isn't big enough!

In practice, we all have to compromise. You may already have a tower and HF beam, in which case you will probably be thinking about mounting a VHF Yagi above it. If you want to be able to operate on both 6m and 4m one perfectly practical solution is to have a duo-band Yagi – many of the European manufacturers now offer several alternative designs according to how large an antenna you can accommodate. Users of the popular range of SteppIR HF Yagis will also be aware that a set of parasitic elements is available for some models which creates a gain antenna for 50MHz. The pattern and gain are not on a par with a good monoband Yagi but it is certainly an easy way of getting started on the band if you have such an HF antenna. The author's own 3-element SteppIR has parasitic elements for both 6m and 4m, although the latter is believed to be a one-off which was installed by the previous owner.

In contrast, for those who want to operate just on 6m and 4m, one effective solution adopted by serious DXers is to stack two 6m Yagis (as this is the band on which maximum gain is likely to be required), mounting a 4m Yagi between the two.

It is worth going back for a moment to talk about arrival angles and what that means for antenna height and design. Single hop Sporadic E signals will almost certainly be arriving at relatively high angles whereas long-haul DX will require very low take-off and arrival angles. Thus, a low antenna may be adequate for the former, whereas a high antenna (with a long boom and multiple elements which tends to compress the vertical lobes) will be preferable for the latter. Stacking will bring down the vertical lobe even more. Height is less of an issue if, fortuitously, you have falling ground in the key directions. A take-off straight over the sea is even better, but few of us are fortunate to live that close to the coast (and, in any case, it will only help in that direction). In practice, the minimum requirement for a reasonably serious 6m DXer with no aspirations to EME is probably a 5- or 6-element long-boom Yagi. That is not to say that DX won't be worked with something less – when the transatlantic path is well-open on multi-hop Sporadic E signals can, at times, be very loud indeed. But, equally, there will be expeditions and DX openings where the signals rarely climb much above the noise level and every dB counts.

What the foregoing discussion makes clear is that there is no single solution as the choice will be determined by your operating interests, your existing installation, your home, garden and immediate surroundings and, no doubt, other factors too, such as planning (zoning) considerations and maybe even compromises to keep your own family happy.

SO WHAT DO I DO?

Don't be put off. In reality most 6m or 4m operators will start with something quite modest, perhaps a 3- or 4-element Yagi for one or both bands, start to enjoy making contacts and then, human nature being what it is, think about something bigger and better for the next DX season. But it doesn't all have to be a case of 'suck it and see'. The good news is that there is plenty of software available nowadays for antenna modelling, not just of the antenna itself but of its surroundings, so that you can determine aspects such as the optimum or the impact of other existing antennas (and hence how much clearance should be allowed when installing on the same mast). It really is worth investing a modest amount in such a modelling program before spending large amounts of money on an antenna which may not achieve what you want. Although most manufacturers nowadays quote antenna specifications in ways which can be compared (this hasn't always been the case), their figures will inevitably be presented in a way which shows their antenna in the best light and which will be based on modelling or measurements in the clear and certainly not take into account your local circumstances, with trees, adjacent buildings and whatever else.

For many years this author used his second tower for 160m in the winter and 6m in the summer, which worked reasonably well and might well do so for others. In that case, though, it would have been nice to have a second, smaller 6m antenna elsewhere in the garden to cover those 'out of season' 6m openings.

There is also an argument for a broadband (perhaps a log-periodic) antenna covering the 30 – 70MHz spectrum to feed your SDR or other general coverage receiver, in order to monitor frequencies outside the amateur bands themselves, while watching the MUF increase. Justin, G0KSC, of InnovAntennas has apparently been seeing increasing demand for broadband antennas for exactly this purpose in recent years, perhaps as SDR receivers become more commonplace. An alternative, if you want to try 4m as well as 6m, is a dual-band Yagi, such as that pictured below.

For those who are really serious about 6m DXing in particular (it is less important for 4m, as there is very little long-haul DX), the next step up is to stack antennas. This is covered in some detail in the next section and there are many references available on the Internet and elsewhere if you decide to follow this

A 6m and 4m dual-band Yagi, with 4-elements on each band on a 2.1m-long boom.

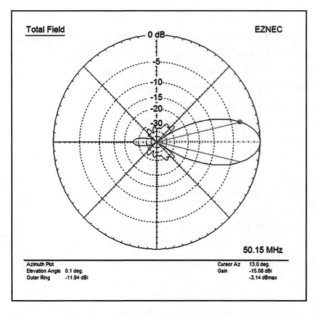

Fig 3.1: *EZNEC* plot of G3WOS's bayed pair of 7-element 6m LFA Yagis, showing the clean pattern with minimal side lobes.

route. Stacking helps to compress the vertical lobe of your antenna, reducing the take-off angle. By using a suitable switch-box you can also arrange to be able to listen and / or transmit on various combinations of the two antennas (both in-phase, both out of phase, top antenna only, bottom antenna only). Until you have tried this capability, it is hard to understand just how useful it can be.

There are various ways in which antennas can be stacked. Kerry, G8VR, has a stack of three Yagis which, given his limited real-estate and taking account of other factors made sense for his location compared with, say, a two-stack of larger Yagis. Chris, G3WOS, for his EME array, has moved to a horizontal stack of two long-boom Yagis with elevation control (easier to achieve with the horizontal stack than with a vertical one) – see **Fig 3.1**.

There is no single answer. A *serious* EME array, but only for those with deep pockets and plenty of space, will normally consist of a stacked and bayed array of four long-boom Yagis because this is really the minimum to be able to work a DX station who, at his end, has just a single Yagi.

While few amateurs nowadays would want to build their own equipment, at least for serious DX and contest work, many people still prefer to build their own antennas, even for VHF. Many designs exist in the literature and on the Internet and the InnovAntennas website carries several designs intended for personal use. One problem, in the UK at least, used to be obtaining good quality aluminium in small quantities but there are now several retailers who offer this (search the Internet to find your nearest vendor), with tubing in suitable sizes for nesting and telescoping. Along with the various modelling software available, there really is nothing to stop the serious enthusiast from designing and building an antenna tailored exactly to his needs.

STACKING ANTENNAS

The design of an antenna system should take into account what the antennas are required to do. In the case of 6m DXing, arrival angles of signals vary dramatically. Long-haul multi-hop Sporadic E or F-layer signals can come in at (and need to be transmitted at) very low angles of just a few degrees above the horizontal. Short hop Sporadic E signals can come in at angles of up to about 15 degrees. No single antenna is going to give optimal performance in both situations. Three possibilities present themselves. The first is to have something like a low dipole or short Yagi for close-in work and a second, high long-boom Yagi for the long-haul low-angle signals. This arrangement also has a potential advantage in spotting band openings because the long Yagi will have a very narrow beamwidth and may be pointing in the wrong direction as the band opens up. Equally, for contesting a narrow beamwidth can mean missing possible callers, so to have the two antennas and be able to switch easily and quickly between them could be a handy arrangement.

STACKING YAGIS

Life is simplest when both antennas are identical. The basic arrangement requires both antennas to be fed in-phase, with the power distributed equally between them. There are two ways of achieving this. One is to use a 2-way power splitter, such as those sold by RF Hamdesign [2]. Each antenna is then fed with an equal length of cable from the two output ports of the splitter. The other is to use co-axial cable to make the feed arrangement. This can be done in various ways – it is not sufficient simply to have a 'Y' arrangement of identical types of co-axial cable as this will lead to impedance mismatches. One solution is a completely symmetrical phasing harness using two equal lengths of 50Ω feeder, and then a quarter-wave transformer made from two paralleled lengths of 75Ω coax. The 50Ω cables can be whatever length is physically needed, so long as they are both the same. Then wherever you go in the band, the two Yagis will still be fed in the same phase. The paralleled 75Ω lines are also very easy to make up, taping the cables together and making the joints inside a small pill-box (after testing, fill the box with hot melt glue). The difficulty is that they need to be an electrical quarter-wave long, which requires taking into account the velocity factor of the cable. Published data on velocity factors will only give you a starting point, so the lengths should be cut slightly over-long and then trimmed while checking the results with a

suitable measurement tool (such as the miniVNA or AA-54 swept analysers – see 'Ancillary Equipment' section in Chapter 2). The system works because the two 50Ω feeds to the antennas are in parallel, resulting in a 25Ω feedpoint. The paralleled 75Ω lengths have an impedance of 37.5Ω which, through a quarter-wave matching section will result in an impedance of 56.25 Ω where you connect the feeder back to the shack, a close enough match to 50Ω to be perfectly acceptable (mathematically, the sum is 37.5 x 37.5 / 25). With this arrangement, and using antennas with a 50Ω feedpoint, there will be some mismatches in the system, but the power will be equally divided between the antennas and the SWR of the system as a whole should be acceptable. There is a lot of useful information on the DK7ZB website [3, follow the links on Stacking] where you will also find an alternative approach, also using paralleled 75Ω cables, which takes the same equation in reverse (in this case 37.5 x 37.5 / 50) which says that the impedance on the far side of such an arrangement is 28.1Ω, so DK7ZB presents designs of 6m Yagis with a 28Ω feed impedance. This is fine if you wish to construct your own antennas, but most (all?) commercial 6m antennas are designed for 50Ω. You can find information on stacking in a number of other places, for example in QST [4], in *Six News* [5, and elsewhere] and on GM3SEK's webpage [6].

Monumental stack of six 6m LFA Yagis at W7EW in Oregon, USA.

A second possibility is to have an elevation rotator similar to those used by satellite enthusiasts. But a 6m Yagi is much larger than most satellite antennas, so most 6m DXers will consider this option impractical.

The third solution, much favoured by those with the wherewithal to do it, is to erect a stack of two Yagis. This is not the place to go into details, though the sidebar covers the basics of stacking.

Stacks have a number of benefits when properly designed and installed. Stacking increases the 'capture area' of your antenna system resulting in more gain for a given boomlength. A stack of comparable forward gain to that of a given single Yagi will have a shorter boomlength and wider beamwidth, which can be beneficial (the downside being that the stack is a three-dimensional system and therefore more of a challenge to erect).

But the real benefits start to accrue by installing a switching system which allows you to select either Yagi independently or to use them together either in or out of phase, commonly called a 'BIP-BOP system' (Both-In-Phase, Both-Out-of-Phase). In phase they will have a low radiation angle, equivalent to a single antenna at a height of the midpoint between the Yagis. Out of phase they will have a high radiation angle. The selection of either antenna independently has less of an obvious benefit but some users of stacks have noticed that when static levels are high the lower antenna may be receiving a much lower noise level than the top antenna of the stack and may therefore give better signal-to-noise on receive. Suitable switching systems are described in the popular antenna texts or ready-made systems can be bought from companies such as Array Solutions of the USA (distributed in the UK by Vine Antennas Ltd, www.vinecom.co.uk).

Whatever antenna system you decide on, you might want, as previously suggested, to model it in software at some stage, to get an idea of how it should be working and how it might be improved. Although the manufacturer will provide gain figures, in reality its performance will be affected by factors such as whether you have it stacked above an HF beam, how high above ground it is, the lie of the surrounding land, and so on. Nowadays all these factors can be modelled if you are so inclined. Popular programs include *Yagi Optimizer* and *EZNEC* (both based on the *NEC* modelling algorithms) and terrain analysis software includes *YT* (bundled with *The ARRL Antenna Book* and designed primarily for HF antennas, but usable for 6m) and *Terrain Analysis* (*TA*). If you model single Yagis over real ground using *TA* or some type of professional modelling software like K6STI's, you will find that they exhibit a deep null somewhere in their vertical pattern within the critical range of arrival angles from 4 to 15 degrees. With stacked Yagis, where either top or bottom can be selected independently, one can choose to place the null at some point other than the optimum arrival angle for a particular signal. Using BIP / BOP / Top / Bottom switching it is almost assured that one of these positions will be optimal for more than 95% of all 6 metre signals you will hear, regardless of propagation.

ANTENNA SUPPORTS

There is nothing unique about antenna supports for VHF but a large 6m antenna and, in particular, a stack are by no means trivial and will require a substantial support. For a stack, the stub mast which supports the array needs to be heavy-walled tubing as the bending moment due to the upper antenna will be substantial. There are many sources of information on masts, towers, winches and the other hardware required to get your antenna into the air. For example, I would particularly recommend GM3WOJ's website [7] for information about the care and maintenance of Versatowers, particularly popular in the UK. The TowerTalk reflector, hosted on www.contesting.com, is also a valuable source of advice and expertise.

THE FEEDER SYSTEM

Having decided on your antenna system, and a suitably robust rotator, do ensure that your budget will also extend to buying good quality feedline. It is no good spending money to get 1dB more gain out of the antenna, for example, while throwing away all that and more in feedline losses. A good rule of thumb is to aim for 1dB or less of feedline loss. Your choice of feedline will then be

Cable type	Loss@ 50MHz	Velocity factor	OD (mm)	Construction notes	Connector compatibility
RG-174	19.2	0.66	2.8	B, PE, T	I
UR76	12.0	0.66	5.0	B, PE. T	I
RG-58	11.0	0.66	5.0	B, PE, S/T	I
Mini-8	8.0 typ	0.75-0.80 typ	6.0 typ	B/BF, F, T	S1
UR67	4.6	0.66	10.3	B, PE, T	I
RG-213	4.3	0.66	10.3	B, PE, T	I
Belden 9913	2.8	0.84	10.3	BF, S, S	S1
Ecoflex 10	2.8	0.86	10.2	BF, S, S	S1
H-100	2.8	0.84	9.8	BF, S, S	S1
Times Microwave LMR-400	2.8	0.85	10.5	BF, F, S	I
Westflex 103	2.7	0.85	10.3	BF, S, S	S1
Andrew LDF2-50	2.4	0.88	9.7	S, F, S	S2
Times Microwave LMR-500	2.3	0.86	12.7	BF, F, S	S
Ecoflex 15	2.0	0.86	14.6	BF, S, S	S2
Times Microwave LMR-600	1.8	0.87	15.0	BF, F, S	S
Andrew LDF4-50	1.6	0.89	16.0	S, F, S	S2
Times Microwave LMR-900	1.2	0.87	17.3	BF, F, S	S
Andrew LDF5-50	0.8	0.89	28.0	S, F, S	S2

Notes:

The top three are not recommended at 50MHz, except for patch cables and very short runs. VK1OD's useful 'RF Transmission Line Loss Calculator' is at http://vk1od.net/calc/tl/tllc.php

Construction notes:

Outer conductor:	B = braided,	BF = braid over foil,	S = semi-rigid copper tube
Dielectric:	PE = solid polyethylene,	S = semi-airspaced polyethylene extrusion,	
	F = foamed (usually polyethylene)		
Centre conductor:	S = solid copper (copper-plated or copper tube),	T = stranded copper (or	
	copper-plated)		

Connector compatibility: I = industry standard for this type of cable,   S = special, S1 = special: may fit a standard connector body but requires different braid clamp and has larger centre conductor,   S2 = requires special connector or major adaptation of a standard connector.

Table 3.1: Matched attenuation of some 50Ω coaxial cables (in dB per 100 metres).

governed by what length of line you have between the shack and the antenna. **Table 3.1** shows the feedline loss for a number of popular types of coaxial cable at 50MHz.

The data in the table was kindly provided by Ian, GM3SEK [8]. It is worth noting that the old US military 'RG' specs have been obsolete for a long time now, so all commercial cables with RG numbers are now 'generics' and may differ in detail from the original MIL specs. 'RG-8' was the first to be made obsolete, and has since spawned so many variants that the designation has become meaningless. In particular, the various kinds of 'Mini-8' bear no resemblance to the original MIL-spec cable, and the type and quality of construction also varies enormously – buyer beware! However, that cable is not normally recommended for use at 50MHz, except perhaps within the shack. The lowest specification of cable that can be recommended for long runs at 50MHz is 'RG-213' (that is the specification that replaced the old RG-8, and is itself now obsolete as well).

Below the line in the table for RG-213, there is a step decrease in attenuation per 100m. From here downwards, we're into the realm of 'serious' low-loss cables.

The next group are all cables of about the same outside diameter as RG-8 / 213 but with a larger centre conductor to reduce loss. This in turn requires a semi-airspaced or foamed dielectric to keep the impedance at 50Ω, but it is the larger centre conductor that's primarily responsible for the lower loss. Within this group, mechanical properties, shielding integrity and cost are the main reasons to choose one cable over another. For outdoor use W3ZZ recommended either Belden 9913F7 or Times Microwave LMR400UF. Both of these are microcellular foam with water impervious jackets.

The cables towards the bottom of the table are much larger in diameter and therefore have lower losses. LDF5-50 is included because it's quite often available surplus at about the same price as LDF4-50 or even cheaper, largely because amateurs perceive it as being difficult to handle. In fact it's hardly any different from 4-50 and connectors are cheaper too, so 5-50 can be a real bargain.

Data used in compiling the table came primarily from the Internet, with sources [9] and [10] being particularly useful and well worth a visit.

If your feedline length is such that the losses will be unacceptable, an alternative approach is to install a masthead preamplifier, as is common on the higher VHF and UHF bands. This results in greater system complexity, as you will need to power the pre-amp and switch it remotely from the shack, and it will inevitably be a potential point of failure. But a *good quality* preamplifier (and I stress good quality, because there is little point in having a transceiver with high dynamic range and intercept point if the preamplifier has neither) will get round those feedline problems. In practice most 6m DXers prefer to spend money on good quality feedline rather than a preamplifier, but a preamplifier is advantageous in certain low signal-level situations. That said, Lance, W7GJ, warns against using a masthead pre-amp in EME installations as he feels it is too likely to be a point of failure. Better, says Lance, to ensure less than 1dB feeder loss and, if necessary, have a pre-amp in the shack.

Chip Angle, N6CA, says in the application notes to his U310 preamp design, "When used in conjunction with good band-pass filters you can bring any good radio up to its best sensitivity without blasting it with 24dB gain from a GaAs FET preamplifier. Your radio won't handle that much gain anyway." Designs for 6m preamplifiers can be found, for example, in the RSGB's *Radio Communication Handbook* and also on various websites (see particularly [11], the N6CA design mentioned above). Commercial models are also available, for example the SSB-SP-6 from SSB Electronics or the LNK-50 from Hamtronics.

REFERENCES

[1] InnovAntennas: www.innovantennas.com

[2] Power splitters: www.rfhamdesign.com/products/powersplitters/index.html

[3] DK7ZB home page: www.mydarc.de/dk7zb/start1.htm

[4] *QST*, January 2008, p82.

[5] *Six News*, Issue 65, May 2000.

[6] GM3SEK on stacking: www.ifwtech.co.uk/g3sek/stacking/stacking2.htm

[7] GM3WOJ on lattice towers: www.qsl.net/gm3woj/latticetower.htm

[8] GM3SEK website: www.ifwtech.co.uk/g3sek

[9] Cable loss data: www.qsl.net/dk3xt/cable.htm

[10] Transmission line loss calculator: http://vk1od.net/calc/tl/tllc.php

[11] U310 Preamp: www.ham-radio.com/n6ca/50MHz/50appnotes/U310.html

4 6m and 4m propagation

WHAT MAKES 6m unique among amateur bands is that it sits at the transition between HF and VHF propagation, so you get the best of both worlds. On the one hand, at the peak of the sunspot cycle you can enjoy superb F-layer propagation, with low absorption. This, coupled with the sort of high-gain antennas that are so much easier to put up for 6m than for the traditional HF bands because of their smaller size, means that it is possible to enjoy long-distance propagation that is at least as good as you may have experienced on 10m at the sunspot peak. Some very long-haul paths have been achieved this way, the UK to Hawaii path quoted in an earlier chapter being a good example.

At the other extreme, serious 6m enthusiasts experiment with moonbounce (Earth-Moon-Earth or 'EME') which, while it works better on the higher bands of

We all like to understand more about propagation: here Costas Fimerelis, SV1DH, presents his findings on the topic of TEP at G3WOS's 2007 6m BBQ.

70cm, 23cm etc is doable on 6m with enough power and a sufficiently sensitive low-noise receiving system. And what has transformed 6m EME has been the availability of the WSJT suite of software, which allows signals to be recovered from well below the noise floor of your receiver.

To get the most out of 6m, you need to familiarise yourself with the various propagation modes and take advantage of each and every one of them. East-west F2 propagation may only be possible for a few years around the peak of each sunspot cycle. Sporadic E (Es, or more correctly E_s) propagation occurs each year in most locations and can be relied on to produce DX contacts even at the bottom of the sunspot cycle, over quite considerable distances, especially in the case of multi-hop Sporadic E or when linked with other propagation modes such as Trans-Equatorial Propagation (TEP).

And it's important to remember, if you want to compete seriously in terms of countries and squares worked, that the major award programmes recognise contacts made by way of digital modes, provided they do not utilise active repeaters (whether ground or satellite based). So a QSO via a repeater would not count for the 6m DXCC award, but a contact made via the moon or by bouncing your signal off the trail of a meteor certainly would.

The following sections describe the main modes of propagation you will encounter on both Six and Four, bearing in mind that F2 propagation is unlikely to be experienced on 4m but all other modes are equally relevant to both bands.

But bear in mind also, when reading the following material, that quite often you may find yourself making QSOs using a combination of more than one propagation mode. For example, at some times of the year stations in Europe or the more northern parts of North America may be able to use Sporadic E to launch their signals into TEP which is happening well to the south of them. In many ways it is discovering how to plug into this sort of phenomenon that makes 6m so interesting compared with the more straightforward propagation modes you may use on other bands. A typical example was during the summer of 2012, when UK stations worked FR4NT, 3B8DB and D64K on 6m, when the solar numbers were modest, to say the least. These contacts were undoubtedly made through a combination of Sporadic E and TEP.

If you learn to recognise characteristics of the different modes you will be able to target your efforts in the directions and at the times most likely to bring results. This is easier to do on Six than on, say, 20m although sometimes the mechanisms at work can be quite involved. For example, one unusual

A combination of Sporadic E and TEP allowed the author to work FR4NT in May 2012 while running about 40W to a SteppIR HF antenna with 6m add-on.

contact was between Scotland and South America where the Scottish station was beaming northwards into an aurora. The backscatter signal from the aurora was then launching into what was probably Es propagation until it hit TEP propagation across the equator!

But perhaps what is even more fascinating is that, even now, we only have a limited understanding of 6m propagation. A classic example is the way in which long-distance paths open up at certain times of the year, for example from Japan to Europe or to the USA, at times of sunspot minima, where the signals are solid, without flutter, for long periods in a way which cannot convincingly be explained by multi-hop Sporadic E. Attempts are being made to explain these phenomena, one mechanism proposed being 'SSSP' (Short Path Summer Solstice Propagation, see later in this chapter). Right now we cannot be certain of exactly what is happening and, in a way, that is the fun of the band. Although we might not know the exact causes of certain band openings, we can certainly enjoy the propagation that they bring.

The rest of this chapter covers the main modes of propagation with which we are concerned in broad terms. For more detailed explanations you will need to go to a specialist propagation textbook or dig out some of the more esoteric research papers that have been written. But it is worth saying once more that even the experts can be mystified at times by propagation on this band. The June 2006 opening between Alaska and Europe, a first for the band, came seemingly from nowhere at pretty much the bottom of the sunspot cycle and in 2007 and 2008, again at the bottom of the cycle, UK and US stations were able to work into Japan when there was no HF propagation at all over that path. Eternal vigilance seems to be the key. Expect the unexpected!

GROUND WAVE

While not a long-distance propagation mode, it is worth understanding the sort of distances that ought to be workable on 6m and 4m without relying on reflections and simply following a direct path. The starting point with radio propagation is that radio waves are just one component of the electromagnetic spectrum. Like other elements of that spectrum (visible light, X-rays), radio waves will travel in straight lines unless something acts to divert them from that path. That something could be a change in density of the medium (think of light waves being refracted as they enter or leave water), or a strong magnetic or electric field.

Even when travelling in a straight line, the waves will not go on for ever. They will be attenuated by the medium through which they are travelling. And they will disperse with distance, depending on the gain of the antenna from which they are propagated. Although a coherent beam of light from a laser, for example, can remain concentrated over extended distances, there is no straightforward way of achieving this with the wavelengths involved in radio frequencies. However high the gain of an antenna, the radiation will spread out with distance.

There is no simple answer as to how far VHF signals will travel, as it will very much depend on your location (hill, valley) and what lies between you and the station you are trying to work. Repeaters will be sited on hilltop locations to allow maximum line-of-sight distances but on both Six and Four, as indeed on other bands, a certain amount of refraction is to be expected, along with possible diffraction over hilltops, to allow distances of several tens of kilometres to be achieved on a reliable basis. So even under flat conditions and without resorting to the more esoteric modes such as moonbounce, you should be able to work other stations in your locality when they are active provided that you don't live in the back of beyond!

The better-equipped your station, in terms of power, antenna height and gain, and receive performance, the greater you will be able to extend your groundwave range. Bear in mind, though, that over a direct path polarisation will not change as it tends to when subjected to the vagaries of the ionosphere, so polarisation needs to be the same at both ends of the path. Repeaters use vertical polarisation because they are designed primarily to extend the range of mobile operators and it is easier to install a vertical antenna on a car than a horizontal one. So if you want to work through a repeater or work mobiles directly you will need a vertical antenna of some sort. Fixed to fixed working tends to be on horizontal antennas, typically (but not restricted to) a Yagi, which will need to be rotatable.

TROPO DUCTING

The troposphere is lower than the ionosphere and tropospheric ducting relies on the effects of weather on the atmosphere, specifically the existence of a *temperature inversion*. This temperature inversion can act, as far as radio waves are concerned, as the wall of a duct along which VHF waves are guided in the same way as in a metallic wave guide.

2m and 70cm operators are very familiar with this effect but it can also be used at 6m and 4m, although signals tend to be weaker and openings less frequent as the duct needs to be larger and larger ducts occur less often.

Tropospheric ducting is also more common in tropical areas and near large bodies of water. Sea paths of 2500km or so are not uncommon in parts of the world (such as Hawaii) where the necessary ingredients come together. The important thing is that the stations at both ends of the path must be able to launch signals into the duct to be able to communicate with each other.

SPORADIC E (Es OR 'E SKIP')

K1SIX, on his website, says *"Regardless of solar conditions, Es remains the bread and butter of the average 6m DXer"*. Sporadic E propagation is caused by reflections from the lower levels of the ionosphere (90 – 130km), giving a normal range of between 700km and 2500km. Double-hop Es, to the eastern Mediterranean from the UK for instance, is quite common and triple- or quadruple-hop contacts to North America occur most years (a great example being G0LCS to VE7XF at 1807UTC on 10 June 2001). At the other end of the scale, when the ionisation is particularly intense, ranges as short as 350 or 400km can be worked.

As a rule of thumb, the more intense the ionisation, the higher the frequencies that will be able to be reflected, and it will become possible to work over shorter paths as the higher angle signals needed for those paths will be reflected back rather than passing straight through the E layer. To take an example that would be typical from the UK, when Sporadic E is present over the centre of Europe and the MUF rises above 50MHz, contacts with Italy, Slovenia etc are likely, but in the first instance nothing will be heard from nearer countries such as Germany, because the ionisation will not be sufficiently intense to reflect the higher-angle signals that would be required to come back down at that shorter distance. But by the time the MUF has reached 144MHz, and 2m Sporadic E contacts with Italy become possible, it is likely that, from the UK, loud German, Polish and similar-distance stations will be workable on 6m. The Italian stations may still be audible – but may well have moved to 144MHz to enjoy the 2m opening.

Obviously, 4m falls between the two – as Sporadic E starts to affect 4m, the skip distance on 6m will have started to draw in.

Es conditions are literally 'sporadic' – they are unpredictable, developing quickly and disappearing again just as quickly. Signals can be (but aren't always) very strong but highly localised, frequently with rapid fading (QSB).

Although it can occur at any time of year, in the northern hemisphere Sporadic E is most likely between May and August, with a secondary peak in December and January. Openings can also take place at any time of day, or even 24 hours a day, but the likelihood is higher mid-morning and around 1800 local time. During a Sporadic E opening the centre of the 'cloud' will usually appear to move from south-east to north-west; this is not so much the cloud itself moving, but rather an effect of the rotation of the earth. **Fig 4.1** (from the G7RAU LiveMUF program)

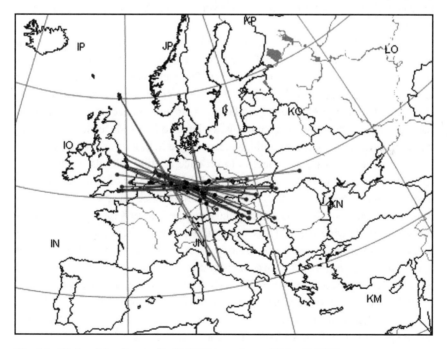

Fig 4.1: Plots of 6m Sporadic E QSOs made on 15 April 2006.

shows Sporadic E QSOs made on 15 April 2006, the first Es of the year as observed by Chris Deacon, G4IFX. Chris comments, *"Because it was early in the year the Es was not particularly widespread and that makes the situation pretty clear, with a single Es cloud over central Germany."*

Particularly for stations well away from the equator, Es accounts for the majority of international contacts during the years of solar minimum, although it's there at solar maximum too, some would say not quite as often. In the peak months of June and July, Es occurs somewhere in Europe and North America on most days. There is invariably a high level of activity as both 6m and 4m operators are very familiar with this form of propagation, and European operators will expect to be able to work many European countries as well as, on 6m, 'across the Pond' to the Caribbean and North America. From the east coast of North America, most states will be workable, along with openings to Europe and North Africa and, on occasions, across the Pacific to Japan.

A high-gain, low take-off angle antenna can be a positive disadvantage when trying to work Sporadic E. The centre of ionisation can move around quite rapidly and arrival angles will typically be quite high. So a small antenna, quite low, can turn out to be the most effective during Sporadic E openings. Perhaps this is another reason why this propagation is so popular, because a lot of DX can be worked with a modest station. This said, the longer-haul paths, launching into a Sporadic E 'cloud' right on the horizon, still demand an effective high-gain low-angle antenna most of the time, especially for double- or multiple-hop Sporadic E where signal strengths are likely to be much lower than with single-hop (every intermediate reflection from the earth, even if over the sea, incurs substantial attenuation).

Much has been written about Sporadic E and its causes and the matter still isn't resolved to everyone's satisfaction. Some argue that the ionisation results from wind shear, while others look elsewhere for an explanation. Some aspects are clear, such as the fact that Sporadic E has a diurnal pattern, peaks near the solstices, and occurs more frequently in latitudes closer to the equator. But none of the explanations that have been investigated have so far produced a 100% correlation or, perhaps more importantly, a means of predicting its occurrence.

Whatever its cause, the effect is to produce drifting 'clouds' of air that is ionised sufficiently to reflect radio signals. When the paths of QSOs are plotted after the event, there is often a clear centre to the reflections, showing that the ionisation is quite localised. But during the major periods of Sporadic E the ionisation can be quite widespread. There is certainly plenty of literature to follow up if you wish to explore further. A good starting point would be the pair of articles that appeared in *QST* in 1997 [1]. One conclusion from the studies reported there is that Sporadic E is more common where the horizontal part of the earth's magnetic field is the largest. This would explain why Sporadic E is most intense in south-east Asia, but infrequent in South Africa, even in summer. Some suggest that, like auroral propagation, Es has a habit of repeating after 27 days (one solar rotation). Whether or not this is true, there is certainly no harm in keeping a weather eye on

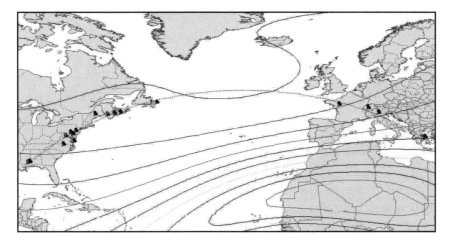

Fig 4.2: Stations worked by multi-hop Es on 4 June 2006 by SV1DH.

the bands 27 days after a good period of Es. A more recent set of articles worth locating is a series of three discussing Sporadic E on 50MHz, 'Extreme-Range 50-MHz Es', that were written by KH6/K6MIO and W3ZZ. They appeared in *CQ VHF* magazine in the Fall 2011, Winter 2012 and Spring 2012 issues.

The best way to get a feel for Sporadic E propagation is to experience it for yourself because it really does have a unique 'feel' quite unlike other propagation mechanisms, with signals often changing from loud to inaudible (or vice versa) in a matter of seconds as the clouds of E-layer ionisation move around. But reports in the magazines also give an idea of what can be worked during such openings, a good example being the report by the late Gene Zimmerman, W3ZZ, in his September 2006 *QST* column of some excellent openings in June of that year between the USA and Japan. Costas, SV1DH, also made a fascinating analysis of an extended multi-hop Es opening on 4 June 2006 across the Atlantic. **Fig 4.2** (from a *DX Atlas* screenshot) shows clusters of stations worked by SV1DH, with the most distant station (in EM50 square) being at 9600km. If the path is assumed to be multi-hop Es, the clusters fit very well to a hop distance of about 1200km, with heard / worked stations at six of the eight landing points, the other two terrestrial reflection points being mid-Atlantic where no stations would be expected. There are some who would argue that Sporadic E is such a variable phenomenon that the probability of getting so many Es clouds to line up in this way is vanishingly low, hence efforts to find other explanations (see SSSP, below), but others remain content that multi-hop Es is a perfectly rational explanation.

SHORT PATH SUMMER SOLSTICE PROPAGATION

Over the past 15 years or so, 6m DXers have noticed that long-distance openings can occur around the summer solstice, even at times of low sunspot activity, with steady signals over a direct path between, say, Western Europe and Japan, or Eastern USA and Japan. This cannot be F2 propagation, as the F-layer MUF is far

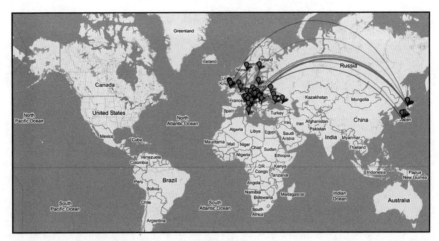

Fig 4.3: The spectacular 6m opening between Europe and Japan on 14 June 2006. (Map from dxers. info)

too low (indeed, from the UK, F2 propagation to Japan has rarely been observed over a direct path, it is almost invariably over a skew path). Multiple-hop Es is the generally-accepted mechanism, but the variable nature of Es propagation and the large number of intermediate hops that would be required (given that the E layer is considerably lower than the F layer) makes it hard to believe that the signals could be so steady and subject to so little attenuation *en route*.

Fig 4.3 shows a spectacular opening between Europe and Japan on the morning (UTC) of 14 June 2006, one of a remarkable sequence of JA–EU openings on successive days. A number of UK stations made QSOs with Japan. Was this by multiple-hop Es or by some new mode of propagation? Strong signals from Scandinavia were audible in the UK at the time and it can be seen from Fig 4.3 that the path from the UK to Japan goes considerably further north than the path from central Europe.

Many 6m DXers have observed the phenomenon and, whatever it is, it may also explain some of the other paths that are observed at that time of the year which, again, can be stable over quite extended periods. Han Higasa, JE1BMJ, who is one of Japan's leading 6m DXers, with a well-equipped station that includes the ability to tilt his array and check arrival angles of signals, has postulated a mechanism which he calls *Short Path Summer Solstice Propaga-*

JE1BMJ's array which can be elevated as well as rotated, allowing a study of arrival angles.

tion (SSSP). He first described his theories in the September 2006 issue of the Japanese magazine *CQ Ham Radio* and, with help from Chris Gare, G3WOS, they were translated into English for a multi-page article which appeared in the UKSMG *Six News*. There isn't space here to do justice to Han's paper, but it is well worth a read and can be obtained by dropping Han or me an e-mail (Han is at higasa@plum.ocn.ne.jp).

While SSSP cannot be predicted, it seems to be most likely when the K index is zero or 1 and the A index is in the lower single digits during the period two weeks before and after the summer solstice. Almost any geomagnetic activity raising the field into the unsettled range will eliminate it. It is also somewhat cyclic like ordinary Es, appearing for a few days and then disappearing for a few days. Perversely, SSSP may become less common as sunspot numbers (and hence geomagnetic activity) increase.

Whether or not you are convinced, Han has started a very useful debate and encouraged 6m DXers around the world to look out for these unusual openings and make contacts, thereby accumulating more data. One day we may know for certain exactly what is happening (and even be able to predict it!) In the meantime, we can continue to enjoy the benefits.

F2 PROPAGATION

HF operators are familiar with F2 as the mainstay of communications on the HF bands. But the Maximum Usable Frequency (MUF) at higher latitudes only reaches as high as 50MHz near solar maximum. It is caused by reflections (or, more strictly, refraction) at a much greater height in the ionosphere (300 – 400km) than Sporadic E, and therefore gives greater range: 3000 – 4000km for a single hop with multiple-hop taking signals out to several times that distance. As the name suggests, this propagation mode relies on refraction through the F2 layer of the ionosphere (the F layer effectively splits into two layers during daytime and recombines into a single layer at night – F2 is the higher of the two layers).

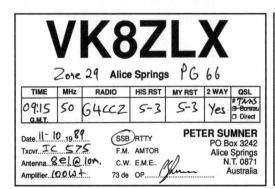

This QSO, between G4CCZ and VK8ZLX, was the first G–VK8 and only the second G–VK QSO on 6m, in 1989, hence the surprise! Such contacts became more commonplace during the following two to three years but F2 propagation during Cycle 24 has been very disappointing.

F2 propagation across the equator at 50MHz is not confined to sunspot maxima and is the basis for Trans-Equatorial Propagation (TEP), described later in this chapter. Stations at higher latitudes can often enjoy the benefits of TEP by launching into it via Sporadic E, one example of the sort of combined propagation mechanism that has already been mentioned. But this section is more concerned with east-west F2 propagation which is the basis for the really long-haul contacts that can be made on Six.

The best of the F2 action takes place around the equinoxes, during February – March and October – November. Signals can be strong but are more often quite weak. They normally vary much more slowly than Sporadic E and openings have a tendency to occur to the same region at the same time of day for several days running. Under good conditions the band can open shortly after daylight, firstly to the east, then following the sun's path across the earth's surface and closing to the west not long after dusk. If you are familiar with 10m propagation, then expect 6m to be similar in terms of the time of day and direction of the openings, but with shorter openings on Six than on Ten.

As a rule of thumb, east-west F2 propagation from higher latitudes (Europe, North America, for example) can be expected on 6m for only about three years around the peak of each 11-year solar cycle, though actual results will depend on the intensity of the cycle (and, in this context, Cycle 24 has been very disappointing and, sadly, current predictions for Cycle 25 are also somewhat pessimistic). The 'solarcycle24' webpage [2] carries up-to-the-minute data including a '6 Meter Tools' tab. North-South F2 propagation is more frequent, benefiting from the higher MUFs in the equatorial zone (which are also responsible for TEP, see next section).

To work over the longest distances via F2, a low take-off angle (probably in the region of 2° to 5°) is recommended, so a long-boom Yagi at a good height (minimum one wavelength, but preferably somewhat more) and in the clear is essential for best results. Sloping ground in the target direction will help your take-off angle, an isolated hilltop location, preferably near the sea (saltwater in the near field also helps) is ideal! Stacking of antennas (covered in Chapter 3) can also bring down the take-off angle.

For a given take-off angle, the effective range of F2 propagation is greater than with E-layer propagation, as the F2 layer is considerably higher. **Fig 4.4** shows the sort of distances that can be expected for one- and two-hop propagation via the E and F2 layers for a range of take-off angles

The same considerations regarding F2 propagation apply to 6m as to the HF bands. For example, polar paths (signals travelling through the auroral oval) are going to suffer from significantly more absorption than east-west and north-south paths, especially when auroral activity (as indicated by the A and K indices) is high.

This is not the place to go into great detail about how F-layer ionospheric propagation works. Essentially, ionospheric propagation depends on the build-up of ions high in the fringes of the atmosphere where wave and particle emissions

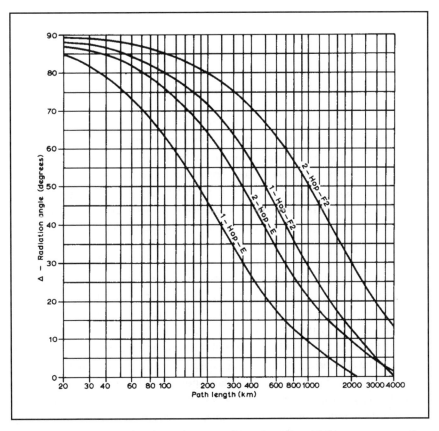

Fig 4.4: The normal distances of one- and two-hop E and F2 layer propagation for a range of take-off angles.

from the sun arrive with sufficient intensity to ionise the air, in other words to split the molecules into positively and negatively charged ions. This layer of ionised air then behaves almost like a mirror as far as radio waves are concerned, causing them to be returned to earth. However, the higher the frequency of those radio waves, the higher their energy, and at any given time there will be a frequency above which all radio waves will have sufficient energy to pass through the ionosphere into space. That frequency will vary according to the level of ionisation. One measure is the Critical Frequency, the highest frequency at which signals which are transmitted vertically will reflect to earth. This will be much lower than the MUF (the maximum usable frequency for signals launched towards the horizon). Thus a Critical Frequency of around 15MHz would probably equate to an MUF of around 50MHz.

The actual level of ionisation will vary according to the amount of energy being received from the sun. And the energy received from the sun varies constantly. There are several cycles involved here: there is the daily effect due to the earth's rotation, there are seasonal effects, just as with the weather, and there is the 11-year sunspot cycle (more accurately, a 22-year cycle, as the polarity reverses every 11 years) superimposed on all those. Disturbances on the sun can also adversely affect ionospheric propagation and these often repeat on a 27-day basis, this being how long it takes for the sun itself to rotate with respect to the earth and present its same face back to us again. Many textbooks explain these effects in great detail as ionospheric propagation has been well studied over the years.

One of the benefits is that there are now many software tools for predicting ionospheric propagation, though their utility for predicting 6m propagation is limited. Any east-west F2 propagation on 6m is likely to be on the fringes of the MUF, dependent on particularly dense patches of ionisation (the ionosphere is never homogeneous) so over a given path the MUF may quite literally be drifting above and below 50MHz in a way which causes signals to appear and disappear in a matter of minutes, or even seconds. Any prediction tools can, at best, simply give an idea of when the MUF on a given path is likely to rise above, say, 30MHz such that it may be worth keeping a weather eye on Six, or when the MUF is somewhat lower and F2 propagation on Six is unlikely in the extreme. But remember that most propagation software was developed originally for commercial use and therefore errs on the side of caution (commercial users are looking for a reasonably reliable path) whereas we as amateurs are happy to settle for the most fleeting of band openings. Also, most such software assumes a typical HF station of, say, 100 watts and a 3-element Yagi, whereas serious 6m operators will have substantially higher ERP (effective radiated power) at their disposal. The trick is to identify those paths where there is a possibility of an opening and then to sit patiently at the radio listening for beacons, DX activity or, perhaps, monitoring commercial frequencies below 50MHz to follow the MUF as it rises. These techniques are described in more detail in the operating chapters of this book.

Solar data (used as the basis for HF propagation predictions) is readily available from a number of sources, giving an indication of solar flux as well as any likely geomagnetic disturbances (the A and K indices) which could adversely affect long-distance propagation, especially through the polar regions. As a rule of thumb, east-west propagation on Six from European and North American latitudes tends to coincide with those periods of the sunspot cycle where the Smoothed Sunspot Number is over 100, or if the solar flux is over 200 for two or more days with a little geomagnetic activity (K = 2 or perhaps 3). But your mileage may vary!

Anyone familiar with 10m propagation knows that some of the tougher paths (e.g. from the UK into the Pacific) can often be worked more easily via long-path propagation than short-path, as the short path would take signals though the northern auroral oval, with significant attenuation. The same is true of Six. High ERP is necessary to exploit these paths, along with low take-off angles and a good (high-gain) receiving antenna. The good news is that 6m antennas are significantly smaller than HF antennas and can be mounted at greater heights in terms of multiples of a wavelength. Stations near the equator enjoy some excellent long-path propagation on 6m at times of sunspot maxima. Costas, SV1DH, reports long-path openings to Japan (30,900km!), with weak signals, but occurring earlier in the sunspot cycle than short-path propagation was noted. Also to Hawaii (26,600km), with no short-path openings observed to KH6. This KH6 path was also observed on a regular basis from Malta during the same period of Cycle 22.

Despite some of the excellent results which have been achieved on Six via F2, it is interesting to note that no G – ZL contact has yet resulted, though this path is relatively common on the 10m band at solar maximum. G – ZL contacts have had to rely on moonbounce.

TRANS-EQUATORIAL PROPAGATION (TEP)

Trans-Equatorial Propagation (TEP) is a variant of F2 which occurs mostly during local afternoon because of two zones of maximum F-layer ionisation which are located just north and south of the geomagnetic equator. **Fig 4.5** is a map of the geomagnetic equator and shows how this follows a somewhat different path to the actual equator, simply because the geomagnetic north and south poles do not coincide with the geographic north and south poles.

'True' TEP is observed when signals are reflected by both of these zones, possibly without a ground-reflection in between. The stations involved must be conjugate, i.e. roughly equal distances either side of the magnetic dip equator, which means that it is only possible to work southern Africa, Indian Ocean or South America by TEP from Europe.

TEP is most common between February and May, and between August and November, often with Es assistance, but can occur at any time of the year, particularly at lower latitudes (closer to the equator). TEP continues even away from sunspot maximum, although it does tend to cover closer distances or disappear at sunspot minimum.

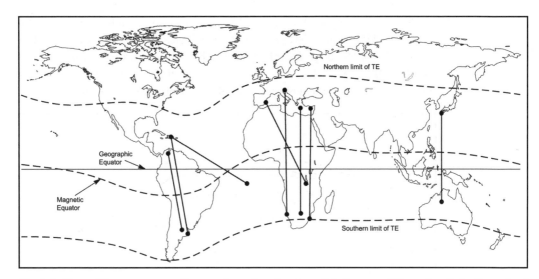

Fig 4.5: Map of the geomagnetic equator, showing how it follows a different path to the actual equator (courtesy of ARRL).

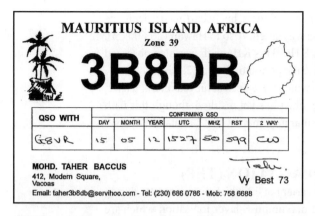

MAURITIUS ISLAND AFRICA
Zone 39

3B8DB

QSO WITH	CONFIRMING QSO						
	DAY	MONTH	YEAR	UTC	MHZ	RST	2 WAY
G8VR	15	05	12	1527	50	599	CW

MOHD. TAHER BACCUS
412, Modern Square,
Vacoas
Email: taher3b8db@servihoo.com - Tel: (230) 686 0786 - Mob: 758 6688

Taher,
Vy Best 73

3B8DB was a surprise new one for many in the summer of 2012, worked from the UK almost certainly by a combination of TEP and Sporadic E.

Because the incidence of F2 drops off rapidly the further north you go, stations in areas such as the Mediterranean and the Caribbean tend to have a better time of it than those of us in places like UK or the northern states of the USA. But, as mentioned previously, quite often mixed-mode propagation, with a single Es hop, for example from the UK down to lower latitudes linking up to TEP from there on, can allow more northerly stations to access the DX especially to Africa and South America, in which case the openings can last well into the evening.

During Cycle 21 Euro-African sector TEP was studied extensively by a dedicated group comprising Ray Cracknell, ZE2JV (G2AHU), and Fred Anderson, ZS6PW, in Africa and Costas Fimerelis, SV1DH, in Athens, taking propagation time delay measurements and observing that it comes in two different modes, the normal 'afternoon mode' (A-TEP) lasting up to 1700UTC and supporting frequencies up to 70MHz with strong, steady and clean signals, and the 'evening mode' (E-TEP) between about 1730 and 2200UTC characterised by weaker propagation and heavily distorted signals up to 432MHz propagating through an ion depletion layer above F2 daylight heights. Propagation drops out completely between the end of one mode and the onset of the other. These times would obviously need to be adjusted for other longitudes, such as the Americas.

Ref [3] gives a good account of TEP, along with some explanatory diagrams. Refs [4] and [5] are the 1981 *QST* articles in which ZE2JV, ZS6PW and SV1DH described their findings regarding TEP, albeit the studies were conducted at 144MHz. But the articles are well worth a read; this is a case where amateurs investigated a propagation phenomenon that was largely unknown in the profes-

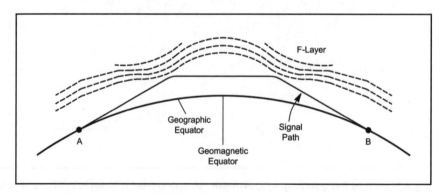

Fig 4.6: Typical cross-section of a trans-equatorial signal path.

sional world and the studies formed the basis of Costas's PhD. An extensive bibliography of 144MHz TEP appears at [6].

Fig 4.6 shows a typical cross-section of a trans-equatorial signal path, showing the effects of ionospheric bulging and a double refraction above the normal MUF.

BACKSCATTER AND SIDESCATTER

Backscatter and Sidescatter are also interesting and useful modes. They occur when a small portion of the signal is reflected, from the Es or F2 ionisation itself or from the ground, either back towards the transmitter (*backscatter*) or to the side (*sidescatter*). Particularly strong scatter can happen when the signal hits a rough sea. Scatter signals have a hollow, echoing sound which is quite distinctive.

Backscatter can give QSOs with stations a few hundred km from you and can be a useful way of filling in missing grid squares, as well as giving a strong indicator of a possible 'normal' opening in the direction from which the scatter comes.

But it's sidescatter that *really* comes up with the goods. The vast majority of the contacts which have been made between the UK and Japan via F2 (as against multi-hop Es) have been by sidescatter from a point somewhere in (or over) the Indian Ocean. The lesson of this, once again, is to make sure that you listen carefully – the best beam heading for those contacts with JA was 90° – 100° rather than the direct path at 30° – 40°. During the fabulous F2 conditions accompanying the secondary peak of Cycle 23, on many days stations in the US worked southern and central Europe direct path at 45° while simultaneously working northern Europe, including Scandinavia and stations as far north as JW and JX, via a skewed path at 140° – 145°. The skewed path signals were very strong, often over S9.

The concept is familiar to HF operators who often find at periods of low sunspots that trans-Atlantic contacts can be made by beaming to the equator somewhere midway along the path. During major contests, when many well-equipped stations are active, this method is particularly effective. On 6m similar considerations apply - signals will be weak but this can be countered by using a high-gain antenna and as much power as you are permitted.

AURORAL PROPAGATION

The natural phenomenon known in Europe as the Northern Lights has been studied and used as a propagation medium by amateurs for more than 50 years. Like the other propagation modes already discussed, auroral propagation is a result of the sun's energy impacting the atmosphere, but in this case it is the effect of a solar flare emitting bursts of energetic charged particles which stream outwards from the sun and spiral towards the earth via the solar wind. These particles are divided by the earth's magnetic field and then follow the field lines to regions known as the auroral zones. These zones are oval shaped and extend outwards from the poles to a radius of 23° on the night side of the earth and to 15° on the daylight side.

Photo: OY1OF

The truly spectacular visual aurora of 5 February 2011 at Tórshavn in the Faroe Islands, photographed by Ólavur Frederiksen, OY1OF.

Visual auroral sightings indicate where the charged particles impinge on the earth's upper atmosphere, ionising the E layer at a height of 110km. The number of auroral openings in any year is dependent on the solar activity. Some areas of the sun can remain active for several weeks, causing repeats of events 26 to 28 days after the initial aurora. This is due to the period of rotation of the sun and events like these are known as solar repeats.

Many such auroral events are able to support the reflection of 6m signals. It is usually quite simple to recognise signals which have been reflected from an auroral zone. They have a rasping sound and the point of reflection will normally be well to the north so that, instead of beaming directly at the station you want to work, you beam at the centre of auroral activity.

Warning signs of impending auroral events

Large increases in the solar noise levels on 95, 136 and 225MHz can be measured during flares, these frequencies being those used by professional observers. However, these increases are also very noticeable on 6m. These sudden iono-spheric disturbances are often followed by short-wave fade-outs on the 14 to 28MHz bands, causing disruption of amateur and commercial HF communications. This is caused by cosmic particles and X-rays ionising the D-layer which absorbs, rather than reflects, radio signals. These particles complete the 150-

million kilometre journey from the flare region on the sun in less than 15 minutes. It should be noted that short-wave fade-outs occur which are *not* accompanied by auroral events, but fade-outs followed by large magnetic disturbances almost always signal an impending aurora.

A good and reliable indication of a forthcoming auroral event is known as *pre-auroral enhancement* and is familiar to HF band operators. This effect is particularly obvious during periods of poor propagation such as in the summer months at sunspot minimum. A typical example occurred on 6 April 1995 when, during a sustained period of mediocre conditions on 14MHz, the band became full of very strong signals from Australia and New Zealand with the propagation akin to that during sunspot cycle peaks. The following day HF band conditions had collapsed and there was a very intense aurora enjoyed by VHF operators on the 50, 144 and even 430MHz bands.

Shortly after the commencement of a magnetic storm the comparatively slow-moving particles ionise the E layers and align along the earth's field lines. VHF radio signals beamed towards the auroral regions are reflected and refracted by the moving area of auroral ionisation. This moving reflector causes frequency shift and spreading, making all auroral signals sound distorted and difficult to copy. Morse signals are transformed into a rough hissing note and SSB voice transmissions vary from a growl to a whisper. The amount of frequency change on signals varies proportionately to the frequency band used. This effect is known as Doppler shift and can be as much as 1.5 to 2kHz HF or LF of the actual transmit frequency on 144MHz. However, the frequency change is less on 6m, making auroral contacts on Six somewhat easier. If the aurora is strong enough, a form of Sporadic E develops, known as Auroral E (see below).

To learn more about what causes radio auroras, you can do no better than turning to *Radio Auroras* by the late Charlie Newton, G2FKZ. It was reprinted by the RSGB in 2012 and updated with new material by Neil Carr, G0JHC [7].

The 2012 edition of Charlie Newton's, G2FKZ, classic book, *Radio Auroras*.

Practical considerations

From the point of view of the 6m or 4m operator, once it is clear that an auroral event is underway (black-out on the HF bands, auroral spots on the *PacketCluster*, information from other sources), it is necessary to beam at the main auroral 'cloud'. This will be broadly to the north but not directly north, so you will need to try various beam headings until signals peak.

It is important to be aware that auroral propagation is a scatter mode. Signals are not refracted as with F-layer propagation but, literally, scattered in all direc-

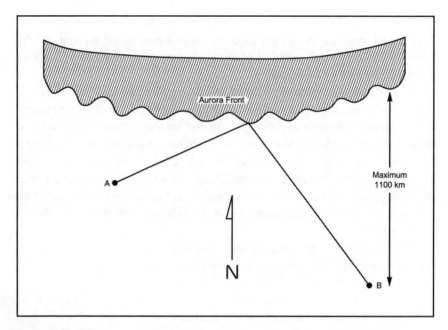

Fig 4.7: Auroral propagation. Point antennas generally north; optimal antenna headings may shift considerably to the east or west depending on the location of the aurora.

tions from regions of intense ionisation. As a result of this scattering signals can be heavily distorted and CW is generally much easier to copy than SSB under these conditions. As mentioned, auroral signals have a characteristic rasping sound quite unlike anything you will experience as a result of other propagation modes, so there can be no doubt of the propagation mechanism involved. **Fig 4.7** illustrates this quite nicely.

Auroral backscatter has two well-defined daily peaks, the first between about 1600 and 2000 local time and the second, during major magnetic storms, around midnight local time. During these events distances up to about 2200km are likely to be workable. The strongest auroras can transform into a condition known as Auroral E, in which distortion levels are lower and path lengths even greater.

It is important to note that while what is generally referred to as 'auroral propagation' is a backscatter mode, Auroral E is forward scatter. The characteristics of the two are therefore rather different: auroral propagation has a rasping sound and the range is generally quite limited; Auroral E produces much cleaner signals and longer ranges. Thus it is not uncommon, for example, for UK stations to hear Doppler-shifted auroral signals from southern Norway and, at the same time, to be hearing clean Auroral E signals from northern Norway.

Predicting auroral events
Before the advent of the packet radio network most operators relied on the telephone to alert others about auroral openings. These links were quite successful.

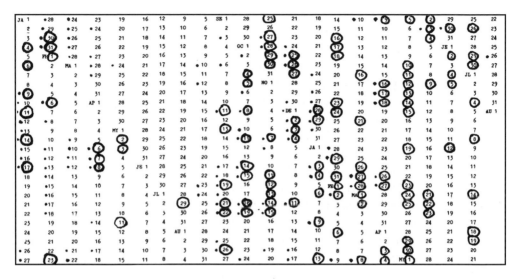

Fig 4.8: A 27-day auroral calendar.

Nowadays serious DXers monitor the *DX Cluster* and the Internet warning systems which provide a real-time record of events as they happen.

Magnetometers are live on the Internet showing you what is and what will be happening, and full daily solar geomagnetic reports are also to be found on the Internet. Many operators keep special 27-day auroral calendars on which they record both the visual and radio events which are reported in *RadCom* and on the *GB2RS* news bulletins. The auroral calendar in **Fig 4.8** which started on 1 January 1977 and records auroral events until 1 August 1978 shows the correlation between the visual and radio events and demonstrates the fact that auroras often repeat 27 days later. A study of Fig 4.8 shows that in the period 1 August 1977 to 1 August 1978 UK observers recorded 70 visual aurora and 100 radio events. These numbers are considerably higher than during the same period for the previous two years, thus reflecting the increase in solar activity as Cycle 21 got under way.

The auroral warning calendar is very easy to use – simply circle any radio events and dot any visual auroras. An operator who was on for the 21 September 1977 event would have been prepared for the repeats which occurred on 18 October, 14 November and 11 December 1977. As the calendar shows, many events repeat in 26 to 28 days, often three or four times, proving that this method can be used successfully. It is also interesting to look at the calendar with hindsight - with the exception of 8 December 1977, the aurora of 18 September repeated consecutively 27 days later seven times. Did we miss an aurora on 8 December or was the event of 4 January and its subsequent repeats unconnected?

Beyond Internet alerting nets, from a practical standpoint one can predict radio auroras to a great extent by looking at certain types of geomagnetic data and comparing these to the time of day these events occur. Thus it is necessary to have both propitious geomagnetic circumstances and favourable geometry to produce effective radio auroras. The geomagnetic values are a K index of 4 or more (www.swpc.noaa.gov/rt_plots/kp_3d.html for the last three days planetary K indices) and a southward (negative) IMF (interplanetary magnetic field) Bz component (from www.swpc.noaa.gov/ace/MAG_SWEPAM_6h.html for the last six hours). In addition the geometry has to be correct. The particles must be travelling earthwards and the geomagnetic storm has to strike at the proper moment during the day, preferably during and early into the afternoon aurora period commencing at 1600 local time. If any of these parameters fails to occur: a negative Bz, a higher than 4 K index, an earthward facing particle stream hitting us mid to late afternoon, the result is likely to be a poor or non-existent radio aurora.

The fact that amateurs note, record and make radio contacts via the aurora is of great interest to professional scientists studying the auroral phenomenon. Predicting correctly the date of an aurora and enjoying the benefits by way of DX is very satisfying. The fact that no two auroral events are ever the same and that it is impossible to predict which countries will be worked simply adds to the attraction of working DX via the aurora.

METEOR SCATTER PROPAGATION

Meteor scatter (MS) is a DX propagation mode which relies on reflecting signals off the brief ionised trail left by a falling meteor. These can last for up to a minute or more on rare occasions, but more usually for fractions of a second. This requires specialist operating techniques, relying nowadays mainly on signal-processing software.

What are meteors? Meteors are particles of rocky and metallic matter ranging in mass from about 10^{-10} to larger than 10kg. About 10^{12} are swept up by the earth each day. At an altitude of about 120km they meet sufficient atmospheric resistance to cause significant heating. At 80km all but the largest are totally ionised, and this ionisation can be used to scatter radio signals in the range 10MHz to 1GHz.

Meteor scatter was once the province of specialist VHF operators but, with the advent of the WSJT software suite, it is now in daily use by many regular 6m and 4m operators with modest stations. Indeed, although meteor scatter propagation has traditionally been associated with the major meteor storms, the earth is constantly bombarded by meteors or other cosmic debris and, using the WSJT software, it is rare that meteor scatter QSOs cannot be made. Signal strengths can be quite high and distances are comparable with those achievable via Sporadic E. The problem lies in the fact that each meteor is very short-lived, so a contact can take a long time, passing just part of the overall exchange as each enhancement occurs. This is where the WSJT software makes life so much easier, as it handles the exchange of information and takes care of the precise timing that is required to effect a meteor scatter QSO. As such the operating technique is quite different to that normally used on the band and is described in detail in Chapter 6.

For the purposes of 6m and 4m operation, there are essentially two scenarios.

Fig 4.9: Seasonal variation of meteor activity.

One is to use regular meteor showers which are predictable and consistent. There are several websites which provide details of forthcoming meteor showers and, more importantly, give data on the timings and the radiant (i.e. the angle of entry into the atmosphere). By beaming at 90° to the radiant (in other words such that the meteors pass across the front of your antenna) you will maximise your chances of making a contact (for reasons which won't be discussed in detail here, the best beam heading is actually offset from this by 9° – 10°: the WSJT software tells you what offset to use). Other stations, to the far side of the meteor shower will be doing the same and, all being well, you will be able to work them by following the techniques described in Chapter 6. **Fig 4.9** shows the seasonal variation of meteor activity, based on a daily relative index.

But debris is constantly entering the earth's atmosphere. The relative velocity of meteors and meteor-like material that meet the earth head-on is increased by the rotational velocity of the earth in its orbit around the sun. Fast meteors strike the morning side of the earth because their velocity adds to the earth's rotational velocity, while the relative velocity of meteors that 'catch up' from behind is reduced (see **Fig 4.10**). This makes dawn the best time of day to try random meteor scatter contacts, though this shouldn't preclude trying at other times of the day.

Fig 4.10: Relative velocity of meteors, as affected by the rotation of the earth.

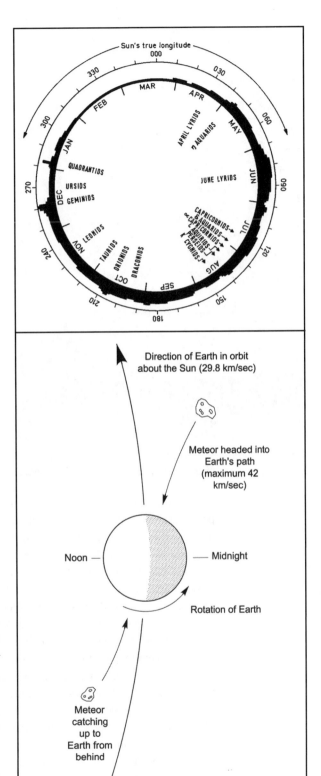

Kerry, G0LCS, and Neil, G0JHC, check out the W6JKV
4 x M-squared EME array at one of Jimmy's 6m BBQs.

MOONBOUNCE (EME)

Not terrestrial propagation, but amateurs have, for many years now, been using the moon as a passive reflector of VHF signals. In theory, any frequency which is not reflected by the ionosphere but passes out into space can be used for moonbounce (also known as EME – Earth-Moon-Earth) communication. In practice, the lowest frequency in regular use is 6m, where moonbounce is possible only when Sporadic E and other enhanced propagation is absent. After all, if the ionosphere is reflecting 50MHz signals back to earth, they cannot be expected to pass through and reach the moon.

Path losses can be calculated with a reasonable degree of accuracy, allowing the station to be designed with sufficient antenna gain, transmit power and receive sensitivity to achieve two-way communications (depending on the station at the other end of the QSO too). Modern software-based systems, discussed in Chapter 6, have made EME accessible to many more amateurs.

Historically, EME required large antenna arrays and it was also necessary to point the antenna accurately at the moon and follow the moon as it moves across the sky. This is no longer the case because, as described in Chapter 6, the WSJT software brings 6m EME into the scope of more modestly-equipped stations. Nevertheless, the station requirements remain quite demanding. Best results are achieved when the moon is just above the horizon, both because this does not require the antenna to be tilted and because your antenna will benefit from a few dB of 'ground gain' at these low angles. Many factors apply in determining the best time for 6m EME contacts, not least because EME requires your signals to pass directly through the atmosphere in both directions whereas, as is clear from this chapter, there are many ways in which those signals can be affected *en route*. EME is not going to work, for example, if a Sporadic E cloud reflects your signals back to earth. At those times you are advised to enjoy the other propagation modes while they exist. But EME is a fallback when other propagation is lacking. At the time of writing, for instance, as mentioned elsewhere in this book, the tough ZL path has *only* been worked from the UK by EME, and a number of recent expeditions have made contacts via EME when other propagation modes have simply not been available.

TROPOSCATTER

Troposcatter propagation utilises the troposphere rather than the ionosphere and takes advantage of irregularities in the troposphere due to differences in the level of moisture, pressure, etc. Such irregularities are always present so troposcatter is available on a daily basis, but signal strengths are low and therefore a high-power station with a high-gain antenna system is required to make good use of this mode of propagation, which can be expected to support contacts up to distances of about 800km.

IONOSCATTER

While on the subject of scatter modes, for sake of completeness it is worth mentioning Ionoscatter. The ionosphere is far from being a homogeneous medium and while the ionosphere as a whole may not be dense enough to reflect 6m signals, occasionally those dishomogeneities may result in denser patches of ionisation sufficient to scatter signals back to earth. This can sometimes happen, for example, in the aftermath of Sporadic E.

It is worth saying that all scatter modes necessarily benefit from good antennas and plenty of power (commercial troposcatter links can use megawatts of ERP) because signals are scattered randomly in many directions so the losses over a given path will be high.

PROPAGATION INDICATORS AND BEACONS

One of the key elements of success on 6m and 4m is to be aware of likely propagation. This chapter has discussed the various types of propagation which occur and given a general idea of when those types of propagation might be expected. But a general idea is not enough. You want to have a very specific indication of when a propagation opening is likely, as even the best forecasting tools can only give some idea of the probability of an opening.

Amateurs have installed beacon transmitters in many locations around the world and these are one of the most useful tools available. Nowadays it is trivial to program their frequencies into the memories of your radio and leave it to scan them on a rotating basis. Lists of 6m and 4m beacons in IARU Region 1 appear in **Appendix D** and **Appendix E**, while G3USF maintains a complete updated list of 6m beacons at www.keele.ac.uk/depts/por/50.htm

As has been mentioned earlier in this book, an even earlier warning is available if you can spot signals from your target area on frequencies below 50MHz, especially if you know of several such transmissions on ever-increasing frequencies, so that you can track the steady rise of the MUF. It isn't possible here to give a list of transmissions that might provide this sort of indication as the list will vary from country to country. In the past VHF TV transmissions were an excellent guide, especially if you had a TV on which to monitor them, as you could then see their ID and know where they were. Most of these have now closed down, so it is usually a case of finding other commercial transmissions in the 40 – 50MHz range in the area in which you are interested. There are plenty of frequency guides, both

NEW PROPAGATION MODES?

by Lance Collister, W7GJ

Hopefully with such popular tools as the ON4KST 6m chat pages, it will be possible to coordinate activities among more 6m hams around the world, and discover new propagation modes that previously have not been noticed. Another development that will help make it possible to discover new propagation modes and band openings is the ever growing popularity and interest in digital modes. For example, the weak signal digital mode of JT65A permits contacts with signals 10dB weaker than CW signals. There very well may be propagation paths that have not been previously discovered by stations running small antennas, barefoot power and CW mode . . . however, as more stations build larger antennas and amplifiers, and experiment with modes such as JT65A, it is quite likely that the extra 25dB gain will really uncover some exciting new possibilities!

in paper form and on the web, which list such transmitters, and over time you will find out which ones work for you and gradually build up a list of them.

FURTHER SOURCES

There is a huge amount of material about VHF radio propagation, both professional and amateur, in books, magazine articles and on the Internet. It is almost impossible to know where to start, though some Internet browsing will give you plenty of pointers (for example, try searching for "50MHz propagation" using *Google* or any other search engine). **Appendix B** lists a number of useful sources. One of the better amateur websites dedicated to VHF propagation is that of DF5AI [8]. But amateurs cannot compete with the professionals for real-time propagation data and you will want to bookmark sites like the excellent spaceweather.com [9] where you can find the latest solar data, auroral data and daily photographs of the sun. You can also sign up for space weather alerts to be e-mailed to you when significant solar events (such as a major flare or aurora) occur.

REFERENCES

[1] 'Sporadic E – A Mystery Solved?' Parts 1 & 2, *QST* October / November 1997.

[2] Solar Cycle 24: www.solarcycle24.com

[3] TEP article: www.ips.gov.au/Category/Educational/Other%20Topics/Radio%20Communication/Transequatorial.pdf

[4] *QST* TEP article, part 1: www.vhfdx.net/docs/qst_te_nov_1981_part1.pdf

[5] *QST* TEP article, part 2: www.vhfdx.net/docs/qst_te_dec_1981_part2.pdf

[6] TEP bibliography: www.dxmaps.com/tepbiblio.html

[7] *Radio Auroras*, Charlie Newton, G2FKZ (SK), and Neil Carr, G0JHC, 2nd (updated) edition, RSGB, 2012

[8] DF5AI: www.df5ai.net

[9] Space Weather: www.spaceweather.com

5 Basic operating

THIS CHAPTER TAKES A LOOK at the basics of operating under commonly experienced propagation conditions. But it also covers how to find DX. Operating is much as it has always been (though the basics sometimes seem to be forgotten in the heat of a pile-up). But what has changed out of all recognition over the last 20 years or so (mainly since the arrival of *PacketCluster* in the late '80s), and continues to do so, is the way in which DX stations are tracked down in order to get around to the serious business of actually putting them in the log.

A WORD ABOUT MODES

If you are reading this book with a view to using 6m and 4m much in the way that you may have been using FM on 2m, that is to say simplex and via repeaters, then fine. There is something in this chapter about how to do so and you will enjoy plenty of activity on the mode. 4m in particular is home to many local FM nets using modified commercial equipment and is a fine band for such activity.

On the other hand, if you want to chase DX, you will need to expand your horizons somewhat. Yes, when Sporadic E is in evidence you will be able to work out to several hundred miles on FM with loud signals and have a lot of fun in the process. But the serious DX will be on SSB and, under marginal conditions, CW (or WSJT, not just for meteor scatter and EME, but for weak-signal terrestrial working too). More about the latter in the next chapter. But the point to emphasise here is that, if you really want to catch all that DX, then you will need to be able to work CW as well as SSB. This can be quite a challenge at first, but is well worth the effort. The good news is that CW decoders are improving all the time and many DXers make their first forays on to CW

Rod, ZL3NW, and Neil, G0JHC, agree that if you really want to catch all the DX on 6m, you need to use CW.

nowadays by using a decoder for receiving and a pre-programmed keyer (or maybe using the PC's keyboard) for transmitting. This set-up won't be effective for long ragchews on CW but can be sufficient to copy a DXpedition station, to call him, see your callsign come back from him with your signal report, and to answer him with a pre-programmed signal report from you. That's all that's needed for a "good QSO". Though hopefully, over time, your own CW skills will improve and eventually you will be able to do away with the aids and send and receive as well as the best of them.

GETTING STARTED

The basics of operating on Six and Four are exactly as they would be on any other band and readers are advised to consult one of the standard texts such as the RSGB or ARRL operating handbooks. But it is worth reiterating some operating advice within these pages, with specific hints and tips as they relate to Six and Four. This chapter covers the basics that you will need to call upon in everyday operation, during Sporadic E and F2 openings, contests etc. The next chapter covers specialist operating, as it relates to EME and meteor scatter, where unique protocols have been devised to cover the nature of these modes and ensure that 'real' contacts take place (with meteor scatter, for example, as will be explained, a contact may take some considerable time due to the random nature of meteor trails, and it is necessary to agree on the timing of transmissions and how to acknowledge receipt of QSO data in a situation where you don't know when the next few seconds of reflections will actually occur).

The 6m band plans for the UK and the US are shown at **Table 5.1 and Table 5.2**. **Table 5.3** is the UK 4 metre band plan, but it should be noted that as more countries gain access to 4m there is an ongoing need to harmonise band planning in a way that wasn't necessary when 4m was almost exclusively a UK band. Indeed, on both bands, band plans vary from country to country, simply because national allocations vary, but the basic structure remains constant. In common with many other amateur bands, the bottom part of the band is reserved for CW operation, then SSB, then other modes including channelised FM.

FM OPERATION

Operation on the channelised parts of the bands is straightforward and will be familiar to anyone who has worked simplex or via repeaters on other VHF or UHF bands. Always listen before transmitting, keep your transmissions relatively short in case others in the area wish to make use of the channel, abide by your licence requirements for identification, and so on. These FM channels are primarily for day-to-day use, over essentially 'line of sight' distances (in the case of repeaters that means line of sight to the repeater). When the band is 'open', via Sporadic E or other long-distance propagation modes, it may be possible to work over greater distances on FM. But at these times it is more appropriate to move down the band and switch to SSB or CW, modes which will cope better with the fluctuating propagation associated with long-distance working. In any case, don't try to

MHz	Necessary Bandwidth	UK Usage
50.000-50.100	500Hz	Telegraphy Only (except for Beacon Project) (Note 2) 50.000-50.030MHz reserved for future Synchronised Beacon Project (Note 2) Region 1: 50.000-50.010; Region 2: 50.010-50.020; Region 3: 50.020-50.030 50.050MHz Future International Centre of Activity 50.090MHz Intercontinental DX Centre of Activity (Note 1)
50.100-50.200	2.7kHz	SSB/Telegraphy - International Preferred 50.100-50.130MHz Intercontinental DX Telegraphy & SSB (Note 1) 50.110MHz Intercontinental DX Centre of Activity 50.130-50.200MHz General International Telegraphy & SSB 50.150MHz International Centre of Activity
50.200-50.300	2.7kHz	SSB/Telegraphy - General Usage 50.285MHz Crossband Centre of Activity
50.300-50.400	2.7kHz	MGM/Narrowband/Telegraphy 50.305MHz PSK Centre of Activity 50.310-50.320MHz EME 50.320-50.380MHz MS
50.400-50.500		Propagation Beacons Only 50.401MHz WSPR beacons +/- 500Hz
50.500-52.000	12.5kHz	All Modes 50.510MHz SSTV (AFSK) 50.520MHz Internet voice gateway (10kHz channels) (IARU common channel) 50.530MHz Internet voice gateway (10kHz channels) (IARU common channel) 50.540MHz Internet voice gateway (10kHz channels) (IARU common channel) 50.550MHz Image/Fax working frequency 50.600MHz RTTY (FSK) 50.620-50.750MHz Digital communications 50.630MHz Digital Voice (DV) calling 50.710-50.890MHz FM/DV Repeater Outputs (10kHz channel spacing) 51.210-51.390MHz FM/DV Repeater Inputs (10kHz channel spacing) (Note 4) 51.410-51.590MHz FM/DV Simplex (Note 3) (Note 4) 51.510MHz FM calling frequency 51.530MHz GB2RS news broadcast and slow Morse 51.650 & 51.750MHz See Note 5 (25kHz aligned) 51.770 & 51.790MHz See Note 5 51.810-51.900MHz. FM/DV Repeater Outputs (IARU aligned channels) 51.910-51.940MHz Internet voice gateways (10kHz channels) 51.950-51.990MHz. FM/DV Repeater Outputs (IARU aligned channels)

Note 1: Only to be used between stations in different continents (not for intra-European QSOs).

Note 2: 50.0-50.1MHz is currently shared with Propagation Beacons. These are due to be migrated by August 2014 to 50.4-50.5MHz, to create more space for Telegraphy and a new Synchronised Beacon Project.

Note 3: 20kHz channel spacing. Channel centre frequencies start at 51.430MHz.

Note 4: Embedded data traffic is allowed with digital voice (DV).

Note 5: May be used for Emergency Communications and Community Events.

LICENCE NOTES: Amateur Service 50.0-51.0MHz – Primary User.
Amateur Service 51.0-52.0MHz – Secondary User: Available on the basis of non-interference to other services (inside or outside the UK).

Table 5.1: UK 6m band plan (as of February 2013).

MHz	USA Usage
50.0 – 50.1	CW, beacons
50.060 – 50.080	beacon subband
50.1 – 50.3	SSB, CW
50.10 – 50.125	DX window
50.125	SSB calling
50.3 – 50.6	All modes
50.6 – 50.8	Nonvoice communications
50.62	Digital (packet) calling
50.8 – 51.0	Radio remote control (20kHz channels)
51.0 – 51.1	Pacific DX window
51.12 – 51.48	Repeater inputs (19 channels)
51.12 – 51.18	Digital repeater inputs
51.5 – 51.6	Simplex (six channels)
51.62 – 51.98	Repeater outputs (19 channels)
51.62 – 51.68	Digital repeater outputs
52.0 – 52.48	Repeater inputs (except as noted; 23 channels)
52.02, 52.04	FM simplex
52.2	TEST PAIR (input)
52.5 – 52.98	Repeater output (except as noted; 23 channels)
52.525	Primary FM simplex
52.54	Secondary FM simplex
52.7	TEST PAIR (output)
53.0 – 53.48	Repeater inputs (except as noted; 19 channels)
53.0	Remote base FM simplex
53.02	Simplex
53.1, 53.2, 53.3, 53.4	Radio remote control
53.5 – 53.98	Repeater outputs (except as noted; 19 channels)
53.5, 53.6, 53.7, 53.8	Radio remote control
53.52, 53.9	Simplex

Table 5.2: US 6m band plan (from ARRL website).

access distant repeaters – repeaters are there to extend the operating range under flat conditions but if you try using them to work 'DX' you will simply deny access to those, such as mobile operators, who are trying to use them for their intended purpose.

Which takes us nicely on to operating in the lower reaches of the band. Obviously some day-to-day working takes place on CW and SSB, but it has to be said that this part of the band really comes to life during a lift in propagation or when there is a contest running. Contest operation will be covered a little later in this chapter and more fully in Chapter 9, but first some thoughts on DX operating.

MHz	Necessary Bandwidth	UK Usage (Note 1)
70.000-70.090	1kHz	Propagation Beacons only
70.090-70.100	1kHz	Personal Beacons 70.090MHz WSPR beacons +/- 500Hz
70.100-70.250	2.7kHz	Narrow Band modes 70.185MHz Cross-band activity centre 70.200MHz CW/SSB calling 70.250MHz MS calling
70.250-70.294	12kHz	All Modes 70.260MHz AM/FM calling 70.270MHz MGM centre of activity
70.294-70.500	12kHz	All modes channelised operations using 12.5kHz spacing 70.3000MHz RTTY/Fax calling/working 70.3125MHz Digital modes 70.3250MHz DX Cluster 70.3375MHz Digital modes 70.3500MHz Internet voice gateway (Note 2) 70.3625MHz Internet voice gateway 70.3750MHz See Note 2 70.3875MHz Internet voice gateway 70.4000MHz See Note 2 70.4125MHz Internet voice gateway 70.4250MHz FM simplex - used by GB2RS news broadcast 70.4375MHz Digital modes (special projects) 70.4500MHz FM calling 70.4625MHz Digital modes 70.4750MHz 70.4875MHz Digital modes

Note 1: Usage by operators in other countries may be influenced by restrictions in their national allocations.

Note 2: May be used for Emergency Communications and Community Events.

LICENCE NOTES: Amateur Service 70.0-70.5MHz Secondary User: 22dBW permitted. Available on the basis of non-interference to other services (inside or outside the UK).

Table 5.3: UK 4m band plan (as of February 2013).

DX OPERATING

The nature of 6m and 4m propagation is that openings can be very short, very focused and often characterised by rapid fading. Often you will first be aware that a DX station is being worked in your part of the world as a result of a 'spot' on the *DX Cluster* system or maybe by hearing a station within ground wave range of you working the DX. You will point your beam in the appropriate direction, tune to the DX frequency and start listening out for the DX station.

Before calling there are certain things you need to establish. These include determining whether the DX station is working 'split', i.e. listening on a frequency other than the one on which he is transmitting. If you are a regular HF band DXer this is a practice with which you will be very familiar. But if your background is largely on VHF, it may be new to you. The basics are covered in the sidebar 'Split Frequency Operation'.

Then listen to his rhythm and to the information he is exchanging. If you have little experience of working DX, it is easy to assume that you should make a long call: "Three Bravo Nine Charlie, Three Bravo Nine Charlie, this is Golf Three X-ray Tango Tango, Golf Three X-ray Tango Tango, over". But this is a bad idea. With rapid fading you may only have a matter of seconds to make a complete contact before his signal fades into oblivion. On the other hand, if he is consistently loud, he will probably have a big pile-up of callers and want to get through them as fast as possible. He already knows his own callsign, so no need to give that, and give yours just once, or twice at the most, before listening to see who he has responded to. Hopefully it will be you, in which case you can respond with his report, "3B9C thanks, 59". Many VHF operators like to give their QTH locator, but only do this if the DX station appears to be giving this information and asking for it from those he is working. Many DX stations will consider it a waste of time, simply reducing the number of contacts they can make during what may be a brief band opening. This information can easily be obtained from the web or from an exchange of QSL cards. And only repeat your own callsign if he has it wrong and you need to make a correction.

OPERATING DURING A SPORADIC E OPENING

Sporadic E openings on Six and Four are relatively common each summer and at some other times of the year, and propagation can exist for several hours at a time. Using Sporadic E to make QSOs is therefore no different to how you would conduct a QSO via any other day-to-day mode of communication.

Avoid using the so-called calling channels – there is no hope of establishing an Es contact and then changing frequency as is normal practice with other propagation. All too often a station fades out during a contact and there is little you can do other than move on or, if it's a rare one you need, wait around in the hope that propagation will return. The usual rules for a contact apply – it is only valid when a confirmed exchange of calls and reports has taken place – if you haven't heard him confirm your report than there is no way you can be sure you are in his log.

Hopefully the propagation will reappear, or maybe you'll just have to wait until

SPLIT FREQUENCY OPERATION

Nowadays, with the huge interest in DX chasing, most DXpeditions and many DX operators choose to operate split frequency. The concept is simple. By transmitting on one frequency and listening on another, callers can hear the DX station clearly, rather than through a mass of other callers, and therefore know when to go ahead with their contact and, more importantly perhaps, when to remain silent while another contact is taking place. And rather than listening on a single frequency, the DXpedition may choose to listen over a range of frequencies, thereby making it easier to pick out callers.

If you hear a DX station or expedition making plenty of contacts, but you can't hear the callers, then the chances are that you are listening to a split-frequency operation. Indeed, before calling any DX station on his own frequency, it's always worth waiting a moment to determine whether he is working split. Otherwise, if you call him co-channel, you risk the wrath of others who have been waiting and will almost certainly inform you of the error of your ways!

If a DX station is working split, what do you do? In the past many amateurs may have had a problem in that their transceiver was only capable of transmitting and receiving on the same frequency. In practice most transceivers came equipped with RIT and XIT (Receiver and Transmitter Incremental Tuning), allowing for a limited degree of split operation (typically up to 10kHz between transmit and receive frequencies). Most modern transceivers go one step further and have two quite separate VFOs, giving total flexibility (though it's worth practising split operation in the peace and quiet of your own shack before you try it on the air and end up pressing the wrong buttons!) Most radios also allow you to check your transmit frequency, which is handy, because as well as listening to the DX station, half the trick can be finding the people he is working, and putting your transmitter on that frequency. The more expensive radios go one step further still, with a second receiver. With that capability, you can listen to the DX station in one ear of your headphones (I never use a loudspeaker for anything other than casual ragchews, and I suspect most DXers are the same) and the pile-up in the other ear. Then you know exactly what is going on.

Always listen to the DX station carefully. Often he will announce his receive frequency, and may also be giving other instructions (such as "UK stations only") which should always be followed carefully to avoid creating unnecessary interference and slowing things down. If no listening frequency is announced, the general rule is to call about 1kHz up on CW, or 5kHz up on SSB. But, again, a few moments' listening should, in any case, quickly allow you to find the callers he is working. Listen for a little longer and you may also determine a pattern. For example, does the DX station always respond to callers on the same frequency or does he, for example, listen a little higher up the band after each contact, finally dropping back down the band and starting the whole process anew? Does he respond to stations only giving part of their callsign or does he, like many DX operators, only respond to callers when he has their full callsign? And so on. Some intelligent listening can pay dividends, compared with simply calling at random. This is how experienced operators running low power can often get through more quickly than less experienced operators running high power (No surprise there, I suppose. Experience counts for more than brute force in most competitive sports and activities).

Unless the DX station is specifically taking 'tail enders', i.e. stations who call as the previous QSO is coming to an end, never call over the top of a contact in progress but wait until the DX station signs and calls "QRZ?" or similar. Otherwise chaos ensues.

The E4X Palestine expedition had great success on 6m, achieving many 'Firsts'.

another day. The cloud of E layer ionisation, which may be relatively small in area, can move very rapidly and also tilt. For example, from the UK, Maltese stations may be very strong one minute then fade out to be replaced by Sicilians which are in turn replaced by Sardinians, all in the space of a few minutes.

CQ calls should be brief, perhaps 10 – 15 seconds, and used with discretion; if an operator is in an area of high activity, any DX station is likely to have many UK stations calling him and be unlikely to reply to a call. Conversely, if an operator is isolated, he may have an opening all to himself and a pile-up of DX may call him. In this case he will want to work as many as possible in the time available, so he will probably just exchange reports with each station maybe giving the locator on every third or fourth contact (the waiting pile-up will have heard it already).

However, always remember that Sporadic E can be very localised and it is possible for a DX station who is S9 to one operator to be inaudible 10km away – and that a CQ call may well pay dividends, especially just as the band is opening. This variability, as discussed in the propagation chapter, can be even more pronounced on multiple-hop Es: even more reason to keep contacts short and sweet. When multi-hop Sporadic E is in evidence (and, for example, it often occurs late in the day European time in the summer, to the Caribbean and North America) signals will be far weaker than with single-hop, so all the foregoing advice is even more relevant. Short, snappy contacts: not only to ensure that your own contact is good before conditions change, but also to allow as many others as possible to make the contact during what might prove to be a short and one-off opening.

AURORAL OPERATING

The beacon stations closest to the auroral zones are usually the first auroral signals to be heard in the south. The Faeroes beacon OY6BEC (50.035MHz), the Swedish beacon SK7SIX (50.027MHz), the Danish beacon OZ7IGY (50.021MHz) or the Finnish beacon OH1SIX (50.025MHz) are all possible indicators of auroral activity in Europe. Similarly the various northerly 4m beacons that now exist.

Auroral events are usually first noticed by amateurs in Norway, Sweden, Finland and the Baltic states, who will be on the band making contacts before the auroral reflections extend to the south. Due to their northerly locations, they see far more visual auroras and participate in more radio events than stations in, say, southern England. It is rare that auroral propagation will extend into southern Europe, but stations in that part of the world benefit more from other propagation modes. This said, amateurs have noticed that the larger the change in geomagnetic activity, the farther south the area of auroral ionisation extends.

If you hear someone calling "CQ aurora" and you have not operated in an auroral opening before, resist the temptation to call indiscriminately, and listen only. There are going to be hundreds of future auroras and this is your chance to learn the entirely new operating techniques required for auroral contacts. First select horizontal polarisation and beam between north and east at 45° (later the centre of propagation may move westwards). Tune the beacon band; it will take some time to get used to the rough sounding keying of the beacons which will be slightly off their usual frequencies due to Doppler shift. Check which beacons are audible. If, for example, you hear one of the German beacons, the aurora is extending at least as far as Germany, and is therefore a large-scale event which will probably last for a few hours and may repeat later in the evening. Turn the beam between north and east on each beacon heard and it will be noticed that different beam headings give peak signals for different countries. Generally the farthest DX is worked with the beam well off a northerly heading.

Next tune the SSB section of the band and try listening to an experienced local station who is working auroral DX. Due to the distortion he will be speaking slowly, using correct phonetics and possibly end-of-transmission tones. Remember that you will hear the local station direct but the DX stations will be replying via the aurora and will be slightly off his frequency.

A typical SSB auroral contact starts like this:

"CQ aurora, CQ Aurora, GM8FFX, Golf Mike Eight Foxtrot Foxtrot X-ray calling CQ aurora . . ." (repeated slowly several times) " and GM8FFX listening." pip (end of transmission tone).

"GM8FFX, GM8FFX, LA2PT calling. Lima Alpha Two Papa Tango, LA2PT calling GM8FFX . . ."

End-of-transmission tones which give a low-frequency pip are very helpful in auroral openings when signals are weak – the tone readily identifies the end of each station's transmissions and could be a 'K' tone or 'pip' tone. Incremental receiver tuning (IRT or RIT) is a 'must' for auroral reception as the amount of Doppler shift often changes in the middle of a contact.

It will soon be seen that SSB contacts are difficult due to the distortion, and contacts tend to be limited to exchanging reports, names and locators. Many amateurs also exchange and log the beam headings used at both ends of the contact as a study of these figures can reveal the particular area of ionised E layer being used.

During weak auroral events SSB operators in England, Northern Ireland, North Wales and Scotland can work each other and operators with better facilities can contact Norway and Sweden. During strong auroral events SSB stations all over the UK and north-west Europe can work each other.

A new operator listening in the CW section of the band will hear a great many rough-sounding hissing CW signals during a strong event. Experienced operators are used to the strange AC-sounding notes and contacts are completed quickly

VOLUNTARY OPERATING CODE OF PRACTICE FOR SIX METRE OPERATORS

(as issued by the UKSMG in conjunction with JAROC, HARDXA, JAROC, SixItaly, DRAA, LABRE-SP and SSA, and reproduced with kind permission of UKSMG)

6m AS A DX BAND: 6m is a DX band just like any other of the amateur radio high frequency DX bands and it, along with other 6m operators, should be treated with respect and tolerance. *Note: Please remember in Europe that French operators are not allowed below 50.200 so local QSOs held just above 50.200 could affect their ability to work DX.*

LOCAL BAND PLAN: Always respect your local band plan. In Europe this is issued by the IARU. *LOCAL QSOs:* Do not cause nuisance and disturbance to other dedicated 6m local and overseas DX operators with local QSOs within the 50.100MHz to 50.130MHz INTERCONTINENTAL SECTION. If you do wish to local rag-chew, it is recommended that you do this above 50.250MHz where interference will be minimised.

LEARN TO LISTEN: Most 6m DXers spend less than 5% of their time transmitting while 95% or more is spent listening and observing changing band conditions and propagation modes. Learn to recognise propagation mode characteristics and when the band is likely to be showing signs of an opening. This will be far more effective than just calling CQ DX at random and *ad infinitum*. *50.100 - 50.130 INTERCONTINENTAL SECTION:* The INTERCONTINENTAL SECTION is a widely accepted concept and should, in principle, be used for Inter-Continental DX QSOs only, especially the 50.110 calling frequency as discussed below. The definition of what constitutes a 'DX' station naturally lies with an individual operator, especially when a particular station within your own Continent constitutes a new country! The 50.100 - 50.130 INTERCONTINENTAL SECTION should only be used for QSOs between stations in different continents or where the station is outside the range of single-hop Es propagation i.e. roughly 2400km or 1500 miles. We would ask you to think carefully before having any intra-European short distance QSOs in the INTERCONTINENTAL SECTION. For those of us in Europe, this is especially important in periods of multiple-hop Es or F2 propagation to avoid burying Inter-Continental DX QSO opportunities under a layer of European QRM.

PLEASE BE SENSIBLE *and avoid local QSOs in the INTERCONTINENTAL SECTION if at all possible!* As the INTERCONTINENTAL SECTION is heavily used, always listen before you call and always ask if the frequency is being used before you transmit (this should be done on any frequency anyway). Just because YOU can't hear anything, it does not mean that the frequency is not occupied or some rare DX is using it. Remember that operating etiquette calls for you to ask if the frequency is occupied BEFORE calling CQ. 50.110.

INTERCONTINENTAL DX CENTRE OF ACTIVITY: The Intercontinental DX Centre of Activity is 50.110MHz. This should be used for long range DX contacts and such contacts should normally be inter-continental in nature. If a local station returns to your CQ, move quickly to an unused frequency above 50.130MHz. Do not use the DX calling channel for testing or for tuning up your radio/antenna.

Do not encourage pile-ups on 110. If you have a successful CQ ensure that you QSY elsewhere in the band. **50.110 CQing: LISTENING** is the first rule of working rare DX on 6m. So think twice before calling CQ on 110. It would be stupid to say that you shouldn't call CQ but please remember that this is a shared frequency so your reputation will be on line if you insist on calling CQ unceasingly every minute of the day or throughout an opening - even if you do say "CQ DX only" or "CQ outside of [my continent] only". The occasional CQ is good as it can discover an unrecognised opening.

If you are a 6m DXer and have been intensely listening for weak exotica for hours on 110 and up pops a CQ caller, rather than ask him rudely to clear off, ask them POLITELY to QSY and TELL THEM WHY OR WHAT YOU ARE HEARING OR LISTENING FOR, and please GIVE YOUR CALLSIGN . Of course, this applies equally well to any frequency on 6m. Most operators are sensible and will do so - probably because they would like to work the DX themselves! Conversely, if you call CQ or are occupying 100 and someone asks you politely to QSY and GIVES YOU A REASON, do so without arguing about the rights of doing so - remember that that you share this resource with thousands of other operators. If you really must call CQ on 110, think twice, listen for five minutes, cross your legs, count to 100, and if the overwhelming desire is still there go ahead and CALL - but keep it short! At the end of the day the choice is yours and yours alone. Don't forget to QSY when successful unless it is inter-continental DX!

QSO TECHNIQUES: Many operators do not take the time to learn how to DX, develop QSO skills and techniques and jump right in. This is not to be recommended as typical 6m propagation does not allow wasting of time during DX QSOs due to the nature of propagation of the band (borderline HF / VHF). Openings could be very short in time duration and DX stations wants to work as many callers as they can during an opening.

Basically, follow the style and take the lead of the DX operator in providing information. Otherwise keep it simple and to the point as there are other stations who are also waiting in line for a QSO with the DX station. Do not waste time in exchanging unnecessary information such as locator codes, names, QTH, equipment, weather and so on. Just exchange your callsigns and confirm your signal reports and move on to allow other DXers to have their QSOs. Leave out all the extra information (such as Maidenhead squares) unless it is requested. Many opportunities to work a DX station are extremely short and if your operating practices prevent others from working the station it will be remembered by those who missed out for a long time. Next time it may be you who misses out.

FREQUENCY CONFLICTS: With the quickly shifting propagation as regularly encountered on 6m, it is quite possible that two stations who have been occupying a frequency for several hours running pile-ups without hearing each other, to suddenly find themselves in a clash. In these circumstances, operators should mutually resolve the situation as quickly as possible to avoid conflict. It should always be remembered that no individual operator 'owns a frequency', even if you have occupied a frequency for several hours.

DX PILE-UP OPERATING: Working and breaking DX pile-ups can be as frustrating experience on 6m as it is on HF. Manners and good operating are very important. You should listen to the DX stations carefully and not continue to call if they request a particular country or prefix to go back to them if that is not you. You should always go back with your complete callsign, give it quickly and give it only once. There is nothing more frustrating and aggravating for others in a pile-up to you to double with the DX station and miss who they going back to. Of course, you should NOT call if you cannot hear the DX station! If a QSO is uncompleted due to QSB or QRM, don't continue to try and complete the QSO to an excessive degree, use your judgment and call back later. It is likely that others are hearing them OK and can complete a QSO. Take the lead from the DX station and don't call back immediately if they are working someone else. The message is simple, try to avoid calling over the top of the DX station - it does you no good and just upsets your fellow DXers.

SPLIT FREQUENCY OPERATION: When a DX station creates a large pile-up of stations all calling him on their own operating frequency (simplex operating) it creates tremendous QRM problems for those calling and the DX station. Under these circumstances, it is recommended that the DX station uses split operating; that is transmitting on one frequency but listening over a range of frequencies above the frequency being used by the DX station. This mode of operating will significantly increase the QSO rate of the DX station.

However, split operating on 6m can cause TREMENDOUS interference with other DX operators who, through no fault of their own, are running a simplex pile-up in the same split-frequency section of the band. To minimise this interference, it is recommend that a maximum split of 10KHz (definitely NOT 100kHz) is used.

DUPLICATE QSOs: It is always tempting to call a rare DX station every time you hear it. This should be avoided as it means that you taking away the opportunity for the DX station to work a new station and give them their first QSO with the DX country. Use your judgment if the DX station is known to be rare! Conversely, a quick call can sometimes be useful if no one else is going back to the DX station to show that there is propagation.

CW OPERATION: CW is probably the major mode of operation on 6m due to the usually weak nature of many real DX openings. Do not call a CW DX stations using SSB as they will not be able to hear you and you will be causing severe interference to other CW DXers trying to work the station. The contrary is true as well, if you cannot break a SSB pile-up using SSB then do not call using CW!

FM QSOs: All FM transmissions should be made above 50.500MHz for the obvious reason that FM is wide band and could wipe out weak DX signals. There is no acceptable reason to transmit FM below 50.500MHz, as there is plenty of spectrum allocated for this purpose.

Note: This Code of Practice may be amended by UKSMG, JAROC and HARDXA from time to time to reflect current licensing conditions and operating practices.

and efficiently. The letter 'A' is added when calling CQ and is also added after the readability and signal strength report, in place of the normal tone reports which are not sent during aurora openings as no signal sounds T9. The best auroral DX is always worked on CW, just as on other propagation modes; CW is easier to copy in weak signal conditions and contacts are therefore completed much faster.

Auroral openings can occur in three separate phases in a single 24-hour period. The first phase can start as early as 1300 local time but usually takes place between 1500 and 1900 local. The second phase can occur between 2100 and 2300 local and a third phase can run from after midnight till 0600. Very few auroral contacts take place around 2000 local and there is often a fade-out between the evening phase and the after-midnight session. Some auroras have no afternoon phase and start in the evening, often continuing again after midnight. Some auroras have no afternoon or evening phase, only starting after midnight. These are almost always weak events, sometimes heralding a larger occurrence on the next day.

Due to the Doppler shift and distortion on signals, auroral contacts can only be made on CW and SSB. High power is not essential but helps greatly in weak events. During strong events almost anyone can participate. Signals reflected from the auroral curtain do not change polarity, and high-gain horizontal antennas give the best results. Operators who are blocked to the south and south-east enjoy auroral openings as they can work stations that are normally unheard due to the obstructions, by beaming well to the north of the direct path.

The amount of Doppler shift is proportional to the frequency band in use. For this reason, auroral signals on the 50 and 70MHz bands are easier to read and have less distortion than on the 144MHz band. VHF operators coming to 6m from 2m will already be familiar with auroral propagation. HF operators may have experienced such propagation on 10 or 12m. But there is no mistaking auroral propagation when you run across it, the sound of signals being so different to any other propagation mode.

OTHER MODES

The WSJT software suite for weak-signal working on 6m is covered in the next chapter, as it is a major topic in its own right. There is nothing to prevent you using other modes on 6m, for example RTTY (teletype) and SSTV (slow scan television) when conditions are suitable, and some operators do exactly that. The use of these modes is well covered in other literature such as the *RSGB Operating Manual*. To set up for them, you will need your PC linked to your radio and suitable software such as *MMTTY* for RTTY and *MMSSTV* for SSTV. Remember, though, to observe the band plan, which has designated calling channels for these alternative modes.

The RSGB's *Amateur Radio Operating Manual* goes into more detail about WSJT, datamodes etc.

CONTEST OPERATING

A short word about contest operating, which is covered in a little more detail in Chapter 9. You may think of yourself as a DXer, not a contester. But the serious DXer will want to participate in 6m and 4m contests, perhaps not to achieve a high score or maybe not even to send in an entry at all (though it will always be welcomed by the organisers as a help to checking other logs). The reason to participate is simple. Contests bring activity on to bands which otherwise might be silent. And that activity can do two things for you. Firstly it may highlight propagation paths that are open and of which you are only aware because of the activity taking place. And, secondly, it may bring on to the band activity from countries, squares, islands, cantons or other 'counters' which you are chasing, especially as contests are often an excuse for some entrants to go out and set up portable stations for the occasion. So do be aware of contest activity and be ready to take part even if you don't consider yourself a contester, *per se*.

FINDING THE DX

A lot has already been said about the types of propagation you will find on Six and Four and how to spot possible propagation opportunities. But there is much more to working DX than simply being around when suitable propagation occurs. The secret is to have access to good information. For example, the various amateur radio magazines and bulletins have always made a point of carrying news about forthcoming operations, as well as wrap-ups of what has been on and worked. Nowadays the problem, if anything, is one of information overload. Your national magazine, such as *RadCom* in the UK and *QST* in the USA, will have a VHF column, carrying useful information, but their lead times mean that breaking news will often arrive too late to be carried. To supplement these sources, DX chasers turn to Internet-based bulletins such as *The Daily DX* [1], a subscription service by W3UR, the *OPDX Bulletin* [2] and *425 DX News* [3] (these latter two are free, but weekly rather than daily). All of these sources carry 6m information. Most expeditions will also put up a website beforehand, with information about frequencies, times of operation, callsign etc. This may well include contact details so that you can contact the DXpedition either directly or via a pilot station, for example to arrange skeds or pass details about such issues as frequencies to be avoided, etc.

But the real revolution has been in real-time information sources. The way you actually make a contact is much as it has always been. The way you find the DX station has changed out of all recognition unless you are one of the (increasingly few) who refuse to turn to external sources and simply enjoy listening to the band (white noise for much of the time!) until something shows up.

Let's look at a few scenarios to show how a serious 6m DXer might ensure that he doesn't miss the current DXpedition (and the same might apply to 4m, as more DX and DXpedition stations start to appear). The first trick is to track the MUF as it starts to rise, whether by F2, Sporadic E or other means. For this, increasingly, serious DXers are turning to wideband SDR receivers able to monitor several frequency segments simultaneously. They will be able to watch, on the

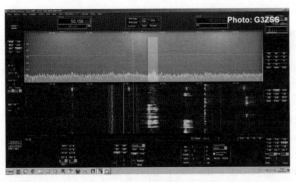

A waterfall display of the 6m band, captured on the display of a Flex 3000 SDR transceiver.

waterfall display, as signals start to appear at ever increasing frequencies. Some will ensure that their wideband receiver is supplemented with a wideband antenna (probably a log periodic) as their 6m antenna will be too narrow in its bandwidth to be an effective antenna across, say, the 30 – 80MHz spectrum.

Then, of course, there are the various Internet-based alerting systems that will let you know when the DX is being heard and worked, albeit not necessarily at your location.

INTERNET-BASED OPERATING AIDS

The most popular source of real-time DX information has to be the Cluster system (formerly *PacketCluster*, but now accessed by most people via the Internet). You are probably already familiar with this global network, whereby amateurs around the world make 'spots' of what they are hearing, this data then being available to everyone connected to the network. The Cluster network also carries solar data (flux, A and K indices) which can be a useful indicator of possible propagation. The Cluster network can be accessed using Telnet (a communications protocol built into your PC's operating system, and also integrated into many logging programs) or you can access it via the *DX Summit* website [4]. *DX Summit* also

The *DX Summit* site, showing 'spots' for the 6m band only.

has a number of other useful features, such as being able to search for 'old' spots so that, for example, if you were out and missed a band opening, you can look back and see what was heard and by whom, which may give you some clues as to best times and frequencies to find the specific station(s) you are chasing.

Although you can set *DX Summit* to show only 6m or 4m 'spots', you may prefer to go instead to a site specifically dedicated to 6m, several of which exist around the world. Equally, you may choose to use the VE7CC Telnet software and select to see only 6m or 4m spots.

The other invaluable facility is the ON4KST chat room [5] (and a few other, similar, chat rooms elsewhere in the world). The ON4KST site has several 'rooms' including one for topband (160m). But the one that will interest readers of this book is that for 6m and 4m. Here you will find a group of like-minded enthusiasts gathered to exchange information, set up JT6M skeds, and generally chat about their favourite band. But the site also shows real-time propagation maps (continually updated from *DX Cluster* spots) and much more, making it a 'must have' tool for 6m stalwarts.

The danger with a site like ON4KST is that there is always a temptation to go beyond simply setting up a sked to the point where you start asking your potential QSO partner such things as "Did you get my report?" or even "Did you get the 57 report I sent you?" which makes any resulting 'QSO' a nonsense. Like any tool, used sensibly it is a valuable aid, but take care not to abuse it.

But this is by no means where the story ends. Nowadays there is a range of tools that draw data from the Cluster or Telnet nodes in a way which is much more useful to you as the needy DXer. For example, most logging programs will take spots from the Cluster network and filter them by DX callsign, country of origin and other algorithms which you, as the user, can define. This means that you only become aware of 'spots' which are of interest to you. Let's say you know (to take the 2012 example) that D64K is active and has a serious 6m enthusiast as part of the expedition team. You might set the filters on your software so that you only see spots for D64K, and only if the station making the spot is in your own CQ zone.

But during the course of a two-week expedition there may be just one or two fleeting openings to such a station (in the case of D64K from the UK that was indeed true although, fortunately, one of them was on a Saturday when many 6m DXers were at home). Can you spare two weeks sat in front of the radio

The August 2012 D64K DXpedition to the Comoros was worked in the UK on 6m.

The *DX Hunter* iPhone app alerts 6m and 4m DXers when a 'new one' appears on the bands.

'just in case'? Probably not. In the past a network of friends might have agreed to alert each other by telephone in such an instance. Later, as technology advanced, some took to carrying pagers to be alerted to DX activity. Nowadays, DXers are much more likely to use a tool such as *DX Hunter* ([6] available for iPhone) which, again, can be programmed for the alerts you need. Then, wherever you are, you can receive that much-needed alert. But what if you are away from home, perhaps at work or at a friend's house? Not to worry, you can even access and operate your station remotely to work the DX while it is there – no need to panic that the band opening will have passed by the time you have driven home. Remote operation is discussed in more detail in the chapter on equipment.

Just to mention a few other facilities available to you, although the range of tools is increasing almost by the day. Firstly, *CW Skimmer* software by Alex Shovkoplyas, VE3NEA [7], will decode CQ calls on CW and present them on screen. If you run the software at home you can monitor the whole of the 6m band (or maybe even 6m and 4m) at the same time, seeing who is active on the band and audible at your location. Many current transceivers have a wideband output before the roofing filter to allow this, as well as to drive external devices such as panoramic adaptors. SDR transceivers such as the FlexRadio range cover 6m directly (those in the Flex 6000 range also include 4m coverage). Again, this has already been discussed in Chapter 2 in the context of equipment selection. But, whether you have a local Skimmer or not, the international Reverse Beacon Network (RBN) [8] allows you to monitor the outputs of Skimmers elsewhere in your own country or farther afield (your choice). You can either sit watching the RBN output, or have it filtered by your logging or other software before sending you an alert to your VHF handheld, Smartphone or other portable device (or even operate an audible alarm around the house should you so choose!)

Having access to all these tools may seem like a form of cheating and there is nothing that says you *have* to use them. But there are many advantages. Look at it, for example, from the point of view of the DXpedition itself. Imagine that someone has spent quite a lot of money to haul equipment to the Caribbean for the summer Es season. The good news is that if and when an opening to, say, Western Europe occurs, then there will actually be people on the band to work who might otherwise have been blissfully unaware and carried on gardening, shopping or whatever else was on their agenda for the day. And such tools have also helped to understand propagation better, as an automated station can, for example, monitor beacons across Africa and elsewhere 24 hours a day, when the chances

de	dx	freq	cq/dx	snr	speed	time
N6EV	W7KNT/B	50062.4	CW BCN	8 dB	15 wpm	0237z 16 Apr
N6EV	W7KNT/B	50062.4	CW BCN	7 dB	15 wpm	0222z 16 Apr
N6EV	W7KNT/B	50062.4	CW BCN	6 dB	15 wpm	0204z 16 Apr
N6EV	W6ZBA	50099.0	CW CQ	16 dB	22 wpm	0133z 16 Apr
N6EV	CE2AWW	50102.0	CW CQ [LoTW]	7 dB	25 wpm	0048z 16 Apr
N6EV	XE2HWB/B	50007.9	CW BCN	4 dB	12 wpm	2337z 15 Apr
EA6VQ	ZD8VHF	50032.9	CW BCN	10 dB	10 wpm	2307z 15 Apr
EA6VQ	ZD7VC	50006.9	CW BCN	9 dB	16 wpm	2243z 15 Apr
N6EV	KI7JA	50102.0	CW DX [LoTW]	8 dB	26 wpm	2238z 15 Apr
N6EV	N6AN	50102.1	CW DX [LoTW]	4 dB	29 wpm	2234z 15 Apr
N6EV	KA7BGR	50075.2	CW BCN	6 dB	12 wpm	2232z 15 Apr
N6EV	K7NV	50101.9	CW DX	4 dB	28 wpm	2226z 15 Apr
N6EV	CE2AWW	50101.9	CW CQ [LoTW]	18 dB	28 wpm	2219z 15 Apr
N6EV	KI7JA	50102.0	CW DX [LoTW]	14 dB	26 wpm	2150z 15 Apr
N6EV	W6SQQ	50089.8	CW DX	4 dB	12 wpm	2143z 15 Apr
N6EV	KA7BGR	50075.2	CW BCN	8 dB	12 wpm	2142z 15 Apr

The Reverse Beacon Network allows users to monitor the outputs of 'CW Skimmers', either locally or wherever the user so chooses.

of someone listening on the right frequency at the right time otherwise are probably very low.

While on the subject of the Internet, in the UK, while there is no 4m club as such, there is an Internet reflector for 4m enthusiasts [9] which is run (at the time of writing) by Sean Williams, M1ECY.

The range of Internet-based tools will undoubtedly continue to proliferate. There are, for example, mapping sites that overlay DX spots on to world maps to show recent 6m propagation, which gives you an almost instant view of, for example, the location of Es clouds. Nowadays there are facilities such as remote receivers through which you can listen to the band from various locations around the world. This can be fascinating in terms of hearing what 6m sounds like from elsewhere but, again, should not be abused (for example, by listening to a DX station that you cannot hear at your own station but raising him from your home simply because you have more transmit power available than he has). It's a matter of ethics, something which has always been relevant to amateur radio, where we conduct our operations from the privacy of our own homes in a way which cannot easily be overseen by third parties (such as those from whom we might want to claim awards). But the Internet has dramatically increased the possibilities for cheating, so the ethics question is more pertinent than it has ever been.

ETHICS AND OPERATION ON SIX AND FOUR

Firstly there are those things which are illegal, either because they contravene our licences or because they contravene the rules of specific awards or contests that we may be interested in pursuing. Then there are those things which are legal but, for various reasons, may be acceptable to some operators but not to others.

Things illegal

Clearly some actions would be quite illegal and to increase your standing in Six metre circles by such conduct would do you no favours at all and, in some cases, could lose you your licence. Examples might include operating with excess power, operating outside your allocated frequencies, or asking someone else to operate your station under your callsign in your absence (allowable in some countries but not in others). Do any of these things and it may increase your score, but would you gain any satisfaction from so doing?

ETHICS

Most handbooks of this type steer clear of the ethics questions, either by choice or by default. But these questions are becoming increasingly relevant and it is worth spending some time thinking about where you stand on some of the issues which are explored in the next few pages.

Then there are actions which, while of no significance to the licensing bodies, are considered out of bounds within the amateur radio community. One example would be submitting forged QSL cards for awards. This is easily done – one friend of mine has quite easily obtained blank QSL cards from over 170 DXCC countries (samples from printers, cards on visitor boards at conventions etc) and it wouldn't be hard to complete these for a hypothetical 6m QSO. Only the most diligent awards manager would spot the fake if the details looked reasonably plausible. There are rumours that some high scores on the 6m DXCC Honor Roll were boosted this way but one has to wonder whether the applicants concerned really get any satisfaction from such behaviour.

But also in this category there are any other forms of behaviour that are against the rules of the awards programme or contest concerned. It gets complicated. A contest sponsor may, for example, prohibit the use of Internet-based support (Cluster, ON4KST chat), but these are quite valid for most awards. So you may decide to operate with such support in order not to miss any rare ones that appear during the contest, but not to send in a contest entry. That's absolutely fine. But as soon as you cross over the line, whether in terms of contest rules or award rules, then, again, there is surely no satisfaction to be gained because your score will not have been gained on the same playing field as other participants.

Matters of choice

For most (hopefully all!) readers the previous couple of paragraphs received your immediate assent. But now we get on to much more grey areas, because many of them come down to personal choice. The classic example is the use of the WSJT suite of programs, which has caused a lot of controversy in recent years. Quite simply, the major awards (let's take DXCC as the prime example) allow contacts made by datamodes to be counted towards your overall totals. So if you aspire to sit near the top of the tables you will need to include those modes in your armoury and be ready to operate meteor scatter, EME etc with all the tools at your disposal.

But for many operators such software is anathema. They consider that it is nothing more than computer talking to computer and is no measure of true operating achievement. "How many squares has your computer worked recently?" is a typical put-down remark.

There is no right or wrong answer. If you have sweated for years to achieve a high score by what might be called traditional means, you will probably be frustrated to see others catching you up at a rate of knots by using these recent advances. But if, for example, you are operating from a somewhat limited urban location, you will probably regard the digimodes as a lifesaver, allowing you to explore propagation modes that, otherwise, would be quite out of your reach.

A slightly different example is the situation discussed elsewhere in this book where your station is remote from your home. Some operators would gain no satisfaction from having to resort to such means to work DX and would only be satisfied with contacts made from home. Others consider it the best way to be competitive on a band which is very sensitive to location. There is no right or wrong answer, though lines have to be drawn. As I write this, for example, the ARRL does not accept contacts made by remote control unless both station and operator are in the same DXCC entity. *CQ* magazine has no such restriction for its awards. But both bodies place restrictions on the location of equipment and, generally, operators for contest operations.

At the end of the day most of these things are a matter of personal choice and you need to decide what matters to you and then, on the whole, stick with it so that you can measure your personal progress in a meaningful way even if you cannot always compare your results with the published results of others.

And finally, while on the subject of ethics and good operating, it is worth

---

DX CODE OF CONDUCT

- I will listen, and listen, and then listen again before calling.
- I will only call if I can copy the DX station properly.
- I will not trust the DX cluster and will be sure of the DX station's call sign before calling.
- I will not interfere with the DX station nor anyone calling and will never tune up on the DX frequency or in the QSX slot.
- I will wait for the DX station to end a contact before I call.
- I will always send my full call sign.
- I will call and then listen for a reasonable interval. I will not call continuously.
- I will not transmit when the DX operator calls another call sign, not mine.
- I will not transmit when the DX operator queries a call sign not like mine.
- I will not transmit when the DX station requests geographic areas other than mine.
- When the DX operator calls me, I will not repeat my call sign unless I think he has copied it incorrectly.
- I will be thankful if and when I do make a contact.
- I will respect my fellow hams and conduct myself so as to earn their respect.

pointing out the the RSGB, UKSMG, many expedition teams and others nowadays highlight their support for the 'DX Code of Conduct' ([10] and reproduced in the sidebar). Given that deliberate QRM, bad operating and similar seem to be on the increase (albeit perhaps more on HF than at VHF) this Code cannot be repeated often enough although, sadly, those who ignore it are probably those who would never actually be aware of it in the first place.

REFERENCES

[1] Daily DX: www.dailydx.com

[2] OPDX Bulletin: www.papays.com/opdx.html

[3] 425 DX News: www.425dxn.org

[4] DX Summit: www.dxsummit.fi

[5] ON4KST: www.on4kst.com/chat/start.php

[6] DX Hunter: www.michiv.de/dxhunter

[7] CW Skimmer: www.dxatlas.com/CwSkimmer

[8] RBN: www.reversebeacon.net

[9] UK 4m reflector: fourmetres@yahoogroups.com

[10] DX Code of Conduct: http://dx-code.org

6 Making weak signal QSOs

T HE 6M BAND IS great fun when there is a good F2 or Es opening, 4m similarly during the Sporadic E season, but much of the time that simply isn't the case. Probably for 90% or more of the time, in fact, with Es present for just a month or two each year (and, even then, only for a few hours each day most of the time) and F2 propagation (on 6m) only at the peak of the sunspot cycle.

What to do at other times? You could ignore Six and Four other than for local chats and stick to HF. Or you could stick with Six and Four and continue to work DX using 'weak signal' modes. Until a few years ago, the only weak signal mode available to amateurs was CW, which has a theoretical advantage of several dB over SSB, providing you use narrow filters to ensure the best signal-to-noise ratio. That is why this book, in several places, recommends that you be ready to use CW when necessary, for example to catch that DXpedition during the short opening when his signal appears above the noise. If this book had been written 10 years or more back, this chapter would have still been about making use of meteor scatter and EME, but the emphasis would very much have been on using CW. It would have talked about the need to squeeze every last dB out of your station to work these modes effectively, but they would have remained very much the do-main of the handful of best-equipped 6m enthusiasts and largely out of reach of the majority. 4m operators could try meteor scatter, but there wouldn't have been many people to work.

Since then the world of weak signal working has changed dramatically. Nowa-days there are other weak signal modes available to amateurs. HF operators will be familiar, for example, with PSK31, which has been around for a number of years now. It allows contacts to take place with low power when, at times, you simply can't hear the station in your headphones although you still have solid copy on the screen.

INTRODUCING WSJT

On VHF, weak signal operating essentially means one thing nowadays: WSJT, the software suite developed by Nobel Laureate Joe Taylor, K1JT (hence the name: Weak Signal by Joe Taylor). No doubt other software solutions could have been

Joe Taylor, K1JT, developer of the WSJT suite of programs (left) with author Don Field, G3XTT, at the International DX Convention in Visalia, California, April 2013.

developed to achieve similar ends, but WSJT is so effective that it has become the *de facto* solution for weak signal working at VHF. The impact of WSJT is so overwhelming that this chapter is essentially nothing more than an introduction to WSJT. It may be that in another five years or so something else will have come along to replace it, though, in reality, what is more likely is that Joe will have released updates to his software to make it even more dominant. In practice, this is exactly what has happened in the years since the *6 Metre Handbook* was published by the RSGB in 2008; the main changes to this chapter are a reflection of how the software has moved on.

WSJT, available since 2001 but progressively added to and upgraded since then (JT6M was released in 2003, with several subsequent updates, MAP65 and ISCAT more recently, for example), allows signals to be recovered from below the ambient noise floor and is generally considered to have a 10 – 15dB advantage over CW, a huge difference when you consider that a linear amplifier typically adds 10dB or so to your signal and an improvement of 10dB in antenna gain requires something like eight times as many elements as you already have (on the basis that doubling the antenna size typically adds 3dB or so)! The impact of that level of improvement is that WSJT brings weak signal working to a much greater number of 6m operators than previously.

Several operating modes are available within WSJT, designed for different types of working. JT65, for example, is optimised for EME working where signals are continuous but extremely weak. JT6M and ISCAT are optimised for 6m meteor scatter working, where signals may be stronger but are intermittent, being reflected off the fleeting ionisation trails which meteors leave as they enter the atmosphere. ISCAT affords the option of either 60 or 30 second sequences, and was designed primarily for marginal ionospheric propagation on 6m. It is not adversely affected by random meteors (as is the case with JT65), so it can be used for paths from meteor scatter range to multihop ionospheric propagation.

But WSJT is by no means confined to EME and meteor scatter working. It can be used with any kind of propagation where signals are too weak to pull out by conventional means. So while you are enjoying S9 contacts around your continent by Es, for example, your neighbour may be working DX several thousand kilometres away, by multiple-hop Es, using WSJT to pull out signals that are inaudible to the ear. Expect many 6m DXers to be using WSJT during the more marginal openings to work DX that SSB and CW operators simply cannot hear.

At the moment JT65A remains the most sensitive mode with WSJT, so it is still the standard used for all 6m EME. Because of its great sensitivity, it also has been used during marginal multihop Es and F2 propagation. It also can greatly extend the range of contacts on ground wave or D-layer scatter propagation, provided there are no meteors to disrupt the decoding. If used over distances less than 1300 miles, it is usually recommended to avoid meteor showers or mornings, when random meteors are most prevalent. For example, W7GJ reports having completed D-layer contacts in the afternoons with stations 600 to 1300 miles away using JT65A mode while both stations were running 100 watts with single Yagis. A contact range that used to be possible only between very large, well equipped stations is now available to very modest sized stations using JT65A. Of course, FSK441 mode is so effective on random meteors, that many modest 6m stations can use that mode for contacts under 1300 miles if they wisely choose the proper time of day to best take advantage of the meteors.

Some amateurs decry WSJT and other digital modes as not being 'proper' amateur radio, in that the PC appears to be doing the work, with minimal intervention from the operator. At the end of the day no-one is going to force you to use WSJT. But it can and does allow 6m enthusiasts to continue enjoying long-distance contacts when more traditional modes aren't up to the job. Which, surely, must be a good thing, if only to maintain band occupancy? From an awards point of view, WSJT contacts are just as valid for most awards (including DXCC) as contacts made via any other mode.

WSJT requires accurate timing of transmissions and contacts are often made by way of a schedule ('sked'), to ensure that both parties are on the same frequency and that the timing of their transmissions is co-ordinated. In the past this may have been organised via the telephone or maybe a QSO on HF. Nowadays it will almost certainly be made via the Internet, for example via the ON4KST chat page, which has been mentioned elsewhere. This is fine and accepted, but no QSO information (signal reports etc) should be passed by that means or the contact becomes invalid. Common sense needs to prevail here. If you and your QSO partner are experimenting, a high level of chat via the Internet may be appropriate to compare notes on how things are going and what, if anything, is being copied. But if you are chasing a contact which you may later need as part of an awards claim, you need to ensure that no accusation can be levelled against you that the contact was in some way compromised by the exchange of key information via the Internet or other extraneous means.

SO JUST WHAT *IS* WSJT?

Developed by K1JT for weak signal working (not strictly true, as mentioned above, as meteor scatter is not necessarily weak signal, but characterised by short 'pings' or bursts of signal between periods of silence) on the VHF bands, the description in the manual says: *"WSJT is a computer program for VHF / UHF communication using state of the art digital techniques. It can decode signals propagated by fraction-of-a-second reflections from meteor trails, as well as steady signals more than 10dB weaker than those needed for conventional CW or SSB"*. WSJT is *Windows* based, and will run on any modern PC. Linux, FreeBSD and Macintosh OS/X downloads are also available. In technical terms, it uses the processing power of a modern PC to integrate weak signals over time in order to be able to extract them from the ambient noise, something which the human ear is not geared up to do.

For 6m enthusiasts the facilities offered by WSJT are as follows:

- JT65, a weak signal mode, designed for EME and extreme troposcatter, and replacing the previous JT44.
- JT6M, primarily for meteor scatter, though suitable also for other propagation types, and optimised for 50MHz. It is no surprise that 6m is the focus for this mode, as 6m is by far the most reliable band for meteor scatter communications, and JT6M has become very popular. JT6M has now been superseded by a newer mode, *ISCAT*, but many, if not most, European 6m operators have stuck with JT6M (unlike in the USA where *ISCAT* has been widely adopted).
- An EME Echo mode for measuring your own echoes from the moon. (This was available in versions up to 4.9.8 and has been reintroduced in v9.)
- FSK441, designed primarily for 144MHz meteor scatter, but used occasionally on 50 and 70MHz.
- MAP65, an addition to JT65 for EME working, designed to merge signals received with orthogonal polarisations to minimise the effects of Faraday rotation.

A detailed description of the WSJT program suite is provided in the online WSJT user guide. The starting point is obviously the main WSJT site of K1JT [1] from which you can download both the software itself and the manual. The website is also a valuable resource for would-be WSJT operators, with much advice on setting up WSJT. UKSMG members may also wish to read an article by Ken Osborne, G4IGO [2], which gives some useful hints and tips on using WSJT.

SETTING UP YOUR STATION FOR WSJT

Your existing 6m station will almost certainly be adequate for meteor scatter operation using JT6M. Indeed, too 'sharp' an array often misses signals coming into the broader capture angle of a smaller single Yagi. Experienced operators tend to favour two or more stacked Yagis since these give extra gain while retaining a broader E-plane lobe. The JT6M website [3] points out that many successful QSOs have been made using less than 50 watts and a 3-element Yagi. That same website has lots of useful tips on interfacing your PC to your transceiver and adjusting the settings for best results. Remember, for example, to turn off speech processing and disconnect your microphone when using digital modes, and keep your power level down as most amateur transceivers are rated for intermittent duty cycle (SSB and CW) whereas datamodes have 100% duty cycle when on transmit.

Frequency setting is of paramount importance. A tolerance of ±500Hz is demanded on CW and 200Hz on SSB. WSJT also requires a high degree of accuracy. A stability of better than 100Hz/h is also expected. Most modern transceivers should be up to the task.

The timing requirements are stringent but not too difficult to meet. As a guide, your PC clock needs to be accurate to much less than a second. There are several software programs that can be used to ensure that the computer's clock is always spot on. Software such as *Dimension 4* [4] is suitable. *Windows XP* time synchronisation is insufficient as it checks the time only once every 24 hours, and your PC clock may drift beyond what is acceptable during that time.

Although slower computers have been shown to work, the recommended minimum computer speed is 800MHz. The computer will also need a serial port or a USB-to-Serial Port Converter, in order to provide automatic computer keying of the transmitter's PTT line. An interface to connect the radio to the computer is typically required. As long as the interface permits operation of PSK31 mode while successfully keying an amplifier, it should work with WSJT.

There are no special requirement as far as the transceiver is concerned. It is helpful if the receiver has an effective noise blanker and the AGC can be turned off (though for JT65 it is usual to turn off the noise blanker and noise reduction as well). It is also essential that the receiver has the capability of a wide (at least 2.8kHz) filter position for USB. The bandwidth requirements vary with the different modes. JT6M, for example, uses about 1.1kHz of spectrum so a 1.8kHz filter would be fine whereas FSK441 requires a filter of at least 2.6kHz bandwidth. JT65 can be used with a 1kHz bandwidth as long as Doppler or your QSO partner's possible TX frequency error does not push the received signal out of this bandwidth. It is not important to have DSP capability in the receiver, since the WSJT software is a digital signal processing package in its own right, working in conjunction with your PC's sound card. As such, it is important when using WSJT to have other sound sources within the computer turned off (e.g. message alerts, music accompanying web pages etc) or these sounds will end up on air. Always remember to disconnect the microphone, too. The problem with any DSP noise reduction is that it can remove the valuable information encoded into the digital messages. Therefore, you should always make sure to turn off any DSP noise blankers or noise reducers. Unless you have 'birdies' or other interfering adjacent signals, it is advisable to keep your USB bandwidth as wide as possible on your particular receiver. Using narrower filter settings introduces loss, and it is always best to let the WSJT program filter out the weak signals at audio level rather than having you introduce attenuation before that point.

While meteor scatter can be achieved with modest power levels, in order to ensure two-way contacts off the moon it is necessary in most cases to use an amplifier. A 400-watt amplifier should be the minimum considered and, typically, a kilowatt amplifier is used. Suitable models have already been mentioned in Chapter 2, but whenever you use a digital mode, and WSJT is no exception, bear in mind that the duty cycle is extremely high because the signal consists of continuous multiple tones. Many amateur radio amplifiers are rated for SSB duty so should be de-rated by 50% or more for digital modes. So, for example, to run the maximum allowable power in the UK, 400 watts at the antenna, you are probably looking at needing an amplifier sold as being capable of 1kW output. Even then, you would be best advised to consult the supplier on its suitability for extended digital transmissions.

GETTING STARTED WITH WSJT

This is not the place to reinvent the wheel. The WSJT user guide and online help will tell you what you need to know and the websites previously mentioned contain a lot of useful hints and tips from those who have been there and done it. Familiarise yourself with the software, test your equipment and practice making

---

WSJT QSO PROCEDURE
(from WSJT Help file)

To optimise your chances of completing a valid QSO using WSJT use the following standard procedures and *do not* exchange pertinent information by other means (e.g. Internet, telephone) while the QSO is in progress!

FSK441 or JT6M: *If you have received:*
…less than both calls from the other station, send both calls
…both calls, send both calls and your signal report
…both calls and signal report, send R and your report
…R plus signal report, send RRR
…RRR, the QSO is complete. However, the other station may not know this, so it is conventional to send 73 to signify that you are done.

JT65: *If you have received:*
…less than both calls, send both calls and your grid locator
…both calls, send both calls, your grid locator and OOO
…both calls and OOO, send RO
…RO, send RRR
…RRR, the QSO is complete. However, the other station may not know this, so it is conventional to send 73 to signify that you are done.

(Sending grid locators is conventional in JT65, but numerical signal reports may be substituted.)

---

dummy QSOs by linking up with others via the JT65 terrestrial chat page [5]. It is important to familiarise yourself with the QSO procedure (see the sidebar) as it is easy to make mistakes in the heat of the moment when you are trying to make your first WSJT contacts.

The rest of this chapter deals mainly with meteor scatter and EME operating, though it also talks about how WSJT can be used at other times when propagation is marginal and the traditional modes are not up to the job of achieving a contact. WSJT may be new to you and you may well be reluctant to give it a try. But the rewards are there to be had and, as has been said, it can help to keep 6m and 4m interesting through those extended periods when other means of communication are useless for anything other than local contacts. Admittedly, EME remains something of a specialist activity, confined at the moment to 6m, requiring plenty of antenna gain, the maximum power your licence allows, and careful attention to station design, especially feeder losses, to minimise overall losses in your system, as signals will almost certainly be below the ambient noise level. So the best thing is to start with meteor scatter and maybe get some new squares and countries under your belt. But who knows, maybe once you are hooked you may be tempted to upgrade your station and give WSJT a try on EME too.

METEOR SCATTER

Meteor Scatter (MS) contacts are made by reflecting signals from the ionised trails of meteors during atmospheric entry (see **Fig 6.1**). MS operation is feasible on both 50MHz and 70MHz (also 28 and 144MHz, but that is outside the remit of this book). Commercial MS links tend to use the low VHF region, as the path availability is greater than at the higher frequencies. The distances which may be covered by typical MS operation are similar to those possible via Sporadic E. Assuming most signals are reflected from a region at an altitude of 110km, the maximum range

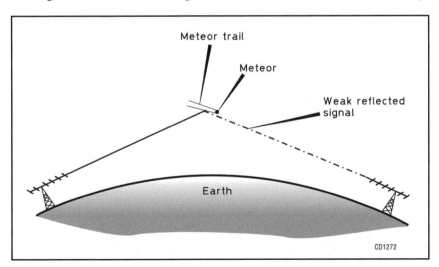

Fig 6.1: Meteor scatter signals are reflected by the ionised trails left by meteors as they enter the atmosphere.

possible with an antenna exhibiting a main lobe at 0° elevation is about 2300km. Typical amateur antennas have a main lobe at 2 – 5° and thus the ranges to be expected are somewhat less.

Most MS contacts on 2m take place during periods of more-or-less predictable meteor activity - the so-called showers. Rather fewer contacts are made via sporadic meteors during the intervals between. In contrast, 6m meteor scatter activity is possible on most days of the year using the specialist JT6M software. SSB and CW contacts via meteor scatter are rare nowadays as JT6M makes life so much easier. Indeed, JT6M has transformed meteor scatter operation on 6m in just a few years, with activity taking place most days.

Table 6.1 lists the major meteor showers, which occur on the same dates each year as the earth passes through those bands of particles on its passage round the sun. Zenith Hourly Rate (ZHR) is the number of meteors per hour at the peak of the shower. This is a calculated maximum number of meteors an ideal observer would see in perfectly clear skies with the shower radiant overhead, based on historical data. For the radio amateur, it gives an idea of which showers are the most prolific. The geographic columns show the QSO path and optimum times. For example, a UK station wanting to work Italy would look at the N-S column to find the best

Shower name	Limits	Max	ZHR	N–S	NE–SW	E–W	SE–NW
Quadrantids	1–5 Jan	3–4 Jan	110	02–06(W) 11–16(E)	11–17(SE)	23–03(S)	00–05(SW) 15–17(S)
April Lyrids	19–25 Apr	22 Apr	15–25	22–02(W) 06–10(E)	23–03(NW) 08–11(SE)	03–06(N)	22–01(SW) 05–08(NE)
Eta Aquarids	1–12 May	3 May	50	03–04(W) 10–11(E)	04–09(NW)	05–11(N)	08–12(NE)
Arietids	30 May–18 Jun	7 Jun	60	04–08(W) 11–15(E)	05–09(NW) 14–16(SE)	08–12(N)	04–06(SW) 10–14(NE)
Zeta Perseids	1–16 Jun	9 Jun	40	05–10(W) 13–17(E)	06–11(NW) 15–17(SE)	09–14(N)	07–07(SW) 11–15(NE)
Perseids	20 Jul–18 Aug	12 Aug	95	23–04(W) 09–13(E)	08–17(SE)	11–01(S)	18–04(SW)
Orionids	16–27 Oct	22 Oct	25	00–03(W) 07–09(E)	00–04(NW)	03–06(N)	05–08(NE)
Taurids S	10 Oct–5 Dec	3 Nov	25	02–05(E) 20–22(W)	20–01(NW)	22–03(N)	00–05(NE)
Geminids	7–15 Dec	13–14 Dec	110	04–09(E) 20–01(W)	22–02(NW) 05–09(SE)	01–04(N) 03–07(S)	03–07(NE) 19–23(SW)
Ursids	17–24 Dec	22 Dec	15	—	07–01(SE)	00–24(S)	16–09(SW)

Table 6.1: Principal meteor showers.

times for a QSO. The letters in brackets show the radiant (direction of arrival). So that, for example, for a N-S QSO during the Quadrantids, the table shows 02–06hrs (W) the radiant being to the west, then from 11-16hrs the radiant is to the east. Both times would be good as the meteor shower is crossing the N-S path.

Serious meteor scatter enthusiasts ensure they are ready for these events, with no distractions in their diaries! Nowadays, though, more casual operators may become aware of meteor scatter opportunities as they watch the *DX Cluster* and see stations exchanging sked information for MS QSOs. This is an example where VHF use of the *Cluster* system differs from HF use. While on HF it is deemed 'unsporting' to arrange skeds or self-spot via the *Cluster* system, on VHF the *Cluster* becomes a useful tool for setting up advance schedules on specialist modes. Of course, for the contacts to be valid no exchanges can be made on the Internet during the contact.

A reread of the meteor scatter paragraphs in Chapter 4 would be useful at this point. As discussed there, the best way of working via a recognised meteor shower is to download data about the shower and to beam at 90° to the radiant. Other stations will also be active during these showers and a CQ on the random MS frequency of 50.230MHz (50.260MHz in North America) may well produce callers. Alternatively, arrange a sked either beforehand or in real-time via the ON4KST chat room [6] or the Ping Jockey [7] websites. General information about forthcoming meteor showers can be obtained from the International Meteor Organisation website [8], but perhaps more useful to amateurs is DL1DBC's meteor scatter site [9] – see the screen shot in **Fig 6.2** – which gives radiant and other data so that you know where to point your beam.

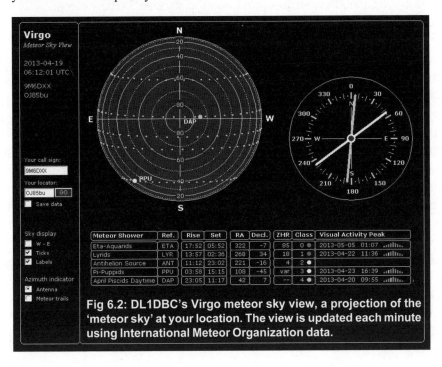

Fig 6.2: DL1DBC's Virgo meteor sky view, a projection of the 'meteor sky' at your location. The view is updated each minute using International Meteor Organization data.

OPERATING PROCEDURES FOR METEOR SCATTER QSOs

Note: These procedures were adopted at the IARU Region 1 Conference in Miskolc-Tapolca (1978), later slightly amended at the IARU Region 1 Conference in Noordwijkerhout (1987), Torremolinos (1990), de Haan (1993) and San Marino (2002). The meeting in Vienna 2004 accepted a slightly reworded version

1. INTRODUCTION

The goal of the procedures described is to enable contacts to be made by meteor scatter (MS) reflection as quickly and easily as possible. Meteor scatter is unlike most other propagation modes, in that neither station can hear the other until an ionised meteor trail exists to scatter or reflect the signals. As the reflections are often of very short duration the normal QSO procedure is not readily applicable and specialised operating techniques must be taken to ensure that a maximum of correct and unmistakable information is received. The two stations have to take turns to transmit and receive information in a defined format, following the procedures as detailed below. Some meteor showers are strong enough to make some of these measures unnecessary but to encourage use of all generally listed showers there is no reason why the suggested procedures should not always be used.

As with operating procedures in general, the virtues of the MS operating procedures are mainly that they are standard and are widely understood throughout IARU Region 1.

2. SCHEDULED and RANDOM CONTACTS

Two types of MS contacts, arranged in different ways, may be distinguished: **a.** A scheduled contact, where two interested stations arrange in advance the frequency, timing and duration of the test, as well as the transmission mode (e.g. Telegraphy, SSB or FSK441) and call signs to be used. Scheduling may be carried out, for example, by exchange of letters or e-mail, by radio via the European VHF Net on 14.345MHz, by Internet chat-rooms or packet radio.

b. A non-scheduled contact, where a station calls CQ or responds to a CQ call. Such contacts are often called "random MS". Random contacts are far more difficult and because you're starting entirely from scratch, it's particularly important for both stations to follow the standard IARU meteor scatter QSO procedures.

3. TIMING

Accurate timing of transmit and receive periods is important for two reasons: to maximise the chances of hearing the other station, and to avoid interference between local stations. The recommended time period for random contacts is, **a.** Telegraphy – 2.5 minute periods, **b.** SSB – 1 minute periods, **c.** FSK441 – 30 second periods.

This practice gives quite satisfactory results. However growing technical standards make it possible to use much shorter periods and amateurs may wish to arrange 1-minute schedules for Telegraphy and shorter periods for SSB especially during major showers. If non-standard time periods are used the first priority is to avoid causing interference to local stations that are using the standard periods.

The recommended standard period for both random and scheduled SSB contacts is 1-minute. However, time periods shorter than this are encouraged during major meteor showers and in North America 15 second periods have been standard for many years either for random or shower meteors. More advanced operators often break half-way though the 15 second period. Quick-break procedures within SSB contacts can be very effective in case the QSO can be completed within one long burst.

Prior to any MS activity it is absolutely vital that clocks need to be set to better than two seconds of standard time. This can be accomplished, for example, by using the telephone 'speakingclock' or GPS time signals. Any clock inaccuracy will result in wasted time and will cause unnecessary interference to other MS stations.

4. TRANSMIT PERIODS

a. All MS operators living in the same area should, as far as possible, agree to transmit simultaneously in order to avoid mutual interference.

b. In Europe, if possible, northbound and westbound transmissions should be made in periods 1, 3, 5 etc counting from the full hour. Southbound and eastbound transmissions should be made in periods 2, 4, 6 etc. North American conventions are reversed – the easternmost station takes periods 2 and 4 in the minute and the westernmost station takes periods 1 and 3.

5. SCHEDULED DURATION

a. Every uninterrupted scheduled period must be considered as a separate trial. This means that it is not permissible to break off and then continue the contact at a later time.

b. Scheduled contacts using Telegraphy or SSB are usually arranged for up to 1-hour duration although during shower periods this can be significantly reduced. Operators using the more efficient FSK441 transmission mode often use 30-minutes or less.

6. FREQUENCIES
Scheduled Contacts:
These contacts may be arranged on any frequency, taking into consideration the mode and band plan. Scheduled contacts should avoid using known popular frequencies and the random MS frequencies.
Non-Scheduled Contacts:
For non-scheduled contacts reference should be made to the relevant IARU Region 1 band plan.

7. QSY FREQUENCIES
To avoid continent-wide interference, which results from a large number of stations attempting to complete contacts on the various MS calling frequencies, a QSY method is recommended. The procedure for moving a beginning QSO off the calling frequency without losing contact is as follows.
a. Telegraphy:
During the CQ the caller indicates on which frequency he/she will listen for a reply and carry out any subsequent QSO. Refer to the relevant band plan for QSY frequencies.

i) Select the frequency to be used for a QSO by checking whether it is clear of traffic and QRM.

ii) In the call, immediately following the letters "CQ", a letter is inserted to indicate the frequency that will be used for reception when the CQ call finishes. This letter indicates the frequency offset from the actual calling frequency used. For instance, CQE would indicate that the operator will listen on the calling frequency +5kHz.

A = 1kHz (CQA) N = 14kHz (CQN)
B = 2kHz (CQB) O = 15kHz (CQO)
C = 3kHz (CQC) P = 16kHz (CQP)
D = 4kHz (CQD) Q = 17kHz (CQQ)
E = 5kHz (CQE) R = 18kHz (CQR)
F = 6kHz (CQF) S = 19kHz (CQS)
G = 7kHz (CQG) T = 20kHz (CQT)
H = 8kHz (CQH) U = 21kHz (CQU)
I = 9kHz (CQI) V = 22kHz (CQV)
J = 10kHz (CQJ) W = 23kHz (CQW)
K = 11kHz (CQK) X = 24kHz (CQX)
L = 12kHz (CQL) Y = 25kHz (CQY)
M = 13kHz (CQM) Z = 26kHz (CQZ)

In all cases the letter used indicates a frequency *higher* than the CQ frequency.

iii) At the end of the transmitting period the receiver should be tuned to the frequency indicated by the letter used in the CQ call.

iv) If a signal is heard on this frequency it may well be a reply from a station who has heard the CQ call and replies on the frequency calculated from the letter used during this call.

v) When the caller receives a signal on the frequency indicated during the call and identifies the reply as an answer on his CQ, the transmitter is moved to the same frequency and the whole QSO procedure takes place there.

b. FSK441:

A similar QSY procedure to that of Telegraphy is used by operators using FSK441 transmissions. However instead of using a letter system, operators should use a number system. Users of FSK441 should indicate the frequency they intend to carry out the QSO by adding the three digits of the nominated frequency. For example CQ383 indicates that the station will listen on 144.383MHz for a subsequent contact.

c. SSB:

The letter system should *not* be used for SSB contacts!

8. QSO PROCEDURE FOR SCHEDULED CONTACTS AND RANDOM OPERATION

a. Calling

The contact starts with one station calling the other, e.g. "G4ASR OH5LK G4ASR ....".

b. Reporting system

The report consists of two numbers:

First number (burst duration)	Second number (signal strength)
2 : up to 5 sec.	6 : up to S3
3 : 5 - 20 sec.	7 : S4 - S5
4 : 20 - 120 sec.	8 : S6 - S7
5 : longer than 120s.	9 : S8 and stronger

c. Reporting procedure

A report is sent when the operator has positive evidence of having received the correspondent's or his own callsign or parts of them.

The report is given as follows: "G4ASR OH5LK 37 37 G4ASR OH5LK 37 37 ....".

The report should be sent between each set of call signs: three times for Telegraphy, twice for SSB and twice for FSK441.

The report must not be changed during a contact even though signal strength or duration might well justify it.

d. Confirmation procedure

i) As soon as either operator copies both callsigns and a report he may start sending a confirmation. This means that all letters and figures have been correctly received.

You are allowed to piece the message together from fragments received over a period of bursts and pings, but it's up to the operator to ensure that it's done correctly and unambiguously.

Confirmation is given by inserting an R before the report: "G4ASR OH5LK R37 R37 OH5LK ...".

A station with an R at the end of the call sign could send "SM7FJE G4ASR RR26 RR26 ...".

ii) When either operator receives a confirmation message, such as "R27', and all required information is complete he must confirm with a string of R's, inserting his own call sign after each eighth R: "RRRRRRRR HG5AIR RRRR ....". When the other operator has received R's the contact is complete and he may respond in the same manner, usually for three periods.

e. Requirements for a complete QSO

Both operators must have copied both callsigns, the report and a confirmation that the other operator has done the same. This confirmation can either be an "R" preceding the report or a string of "RRRR..."'s as explained in paragraph 8.d.ii.

Contacts using SSB are conducted in the same way as Telegraphy or FSK441.

When attempting random contacts, speak the letters clearly, using phonetics where appropriate. It may not be necessary to use phonetics during a scheduled SSB contact, but still speak clearly.

9. MISSING INFORMATION

If a confirmation report (R**) is received it means that the other operator has copied both call signs and the report, yet you may still need something from that station. At that stage, you can try to ask for the information needed by sending a missing information code string.

The following strings may be utilised by operators using Telegraphy to ask for missing information:

BBB.... both callsigns missing
MMM.... my callsign missing
YYY.... your callsign missing
SSS.... duration and signal strength missing
OOO.... all information complete
UUU.... faulty keying or unreadable

The other operator shall respond by sending only the required information. This approach must be used with great caution to prevent confusion.

Note:

These procedures were adopted at the IARU Region 1 Conference in Miskolc-Tapolca (1978), later slightly amended at the IARU Region 1 Conference in Noordwijkerhout (1987), Torremolinos (1990), de Haan (1993), San Marino (2002). Due to significant advances and usage of machine generated modes (such as FSK441) these procedures were updated at the interim meeting, Vienna (2004).

A spectacular meteor trail.

A little forethought is necessary when planning meteor scatter activity. For example, if you are in the UK, you need to catch the meteor showers when they are to the east, as that is where activity will be. There would be opportunities to work to the west later in the day, but there's generally not too much activity from somewhere out in the Atlantic Ocean! It's not the meteor shower that's moving, of course, it's the earth rotating through the meteor belt.

But many 6m and 4m meteor scatter operators don't bother with any of that. Instead they rely on the daily occurrence of random meteors entering the earth's atmosphere. As described in Chapter 4, the peak occurrence of such meteors is around dawn local time, so UK amateurs tend to focus their activity early in the day, perhaps operating from an hour or so before dawn until an hour or so afterwards. Again, most of the activity, for UK-based stations, will be to the east, where it is already later in the day and you are looking for reflections off debris ('rocks' as meteor scatter operators refer to it, despite its microscopic size!) half way between you and the station you are trying to work. With this sort of activity, your beam heading will be towards the station you are trying to work. From other parts of Europe, or within, say, North America, the same logic applies, but you need to work out what this means in terms of whether you are wanting to work to the east or west of your QTH (or, indeed, north or south, in which case local time is the same at both locations).

When arranging a sked on meteor scatter, whether via a shower or at other times, use a frequency other than the random calling channel which is designated for exactly that purpose – random contacts. There is no justification for using that channel when you are making direct arrangements with the other station and, by doing so, you will simply make yourself unpopular with other meteor scatter operators.

The following section talks about meteor scatter procedure. Meteor scatter operating is very different from a 'traditional' QSO. The QSO is pieced together over a period of time, although 6m and 4m contacts can generally be completed much more quickly than is the case on, say, 2m. With WSJT it is possible for an entire contact to be completed during a single 100ms burst, not uncommon in the morning even when there is no meteor shower. On SSB it should be possible to complete a meteor scatter contact within a maximum of about 15 minutes. But the main point is that only once all the data is in place can the contact be considered complete. Thus some very specific guidelines have evolved, which are very different from what you will have encountered on other modes, so they should be studied carefully. They are endorsed and promoted by the IARU to ensure that all operators follow the same procedure. The purpose is to avoid confusion and to ensure absolute clarity by both QSO partners as to whether a valid contact has taken place. The Internet is there for chatting before or afterwards; it should not be used to fill in any missing data!

MS QSO procedure

The intermittent nature of MS propagation means that special operating procedures are necessary. Within IARU Region 1, they are the subject of international agreement (see the sidebar), and thus should be employed. Failure to do so has resulted in lost contacts. This said, the IARU recommendations have been rather overtaken by events in one sense, in that they are effectively built into the WSJT software.

If you do want to use CW, the same guidelines apply. On CW, high speeds are traditionally employed in Europe, though in North America it is more common to use speeds which can be copied by ear. During high-speed CW skeds, speeds from 200 to over 2000 lpm (letters per minute) are in use. In random MS work 800 lpm is the recommended maximum speed. Most operators use memory keyers to send and tape recorders to receive but sophisticated computer software is also available to deal with these speeds. But on 6m, SSB was the more popular 'analogue' (pre-WSJT) mode because information transfer is faster when copied by ear.

In the case of JT6M or ISCAT, users should consult the WSJT documentation for the best way to set up their station and conduct MS QSOs. The jt6m.org website [3], run by G0CHE, is also full of helpful advice on using the mode as well as having a table of meteor showers and some interesting operating tables. The 2008 Perseids Challenge results, for example, give an idea of just what can be worked via meteor scatter – the 2007 Challenge leader was SM7CMV with 223 grids worked (G7CNF led the 4m table, with 69 grids). The 2010 Geminids 6m Challenge was won by S57TW with 80 grids.

Arranging skeds

Once the procedures used for MS have been understood, and the necessary equipment gathered together, the intending MS operator has to find someone to work. There are really two choices: to make a first contact directly on the frequencies set aside for random MS operation or to arrange a sked with an active MS station. The calling frequency is 50.230MHz (50.260MHz in North America) but this can quickly become busy during major showers, so be prepared to move away and, when arranging skeds, as already mentioned use a frequency other than the random calling channel.

When arranging an MS sked, a certain minimum amount of data needs to be exchanged:

a. The date

b. The times of the start and finish (two hours maximum duration is usual – if the contact hasn't been completed in this time leave it and try another day)

c. Length of transmit and receive periods

d. The frequency

e. Which station transmits first

f. (For CW skeds) CW speed.

But nowadays, with the ubiquity of the Internet, many meteor scatter skeds are made 'on the fly' with others who are connected to the chat room at the same time, so some of the above data is implicit.

Finally, another site which can be highly recommended is that by OH5IY [10] which features lots of information about meteor scatter information and some downloadable software.

MOONBOUNCE

(This section compiled with substantial help from Lance, W7GJ, one of the leading operators on 6m EME and has been updated with assistance from Chris, G3WOS, who is a leading UK proponent of 6m EME.)

Probably the most challenging type of VHF / UHF DX communication is that using the moon as a passive reflector, popularly known as EME (earth-moon-earth), or 'moonbounce'. The first known use of the moon as a relay was by the United States Navy which set up a circuit between Washington DC and Hawaii using 400MW (400,000kW!) ERP although the excellent *Wikipedia* entry on Moonbounce [11] suggests that the idea was first mooted by W J Bray of the British Post Office in 1940.

Photo: NASA

The first *amateur* EME contact was on 1296MHz in July 1960 between W6HB and W1BU. Back then CW was the main operating mode and, until the advent of WSJT software, EME on 6m was accessible only to those with very large stations (plenty of power and multiple, stacked antennas).

In his VHF column in the January 2000 issue of *CQ Magazine*, Joe Lynch N6CL, wrote, "*As the F2 propagation waned, those of us perilously close to the magic number 100 started to scramble to see how else we could make those last few contacts to reach the goal. For a while EME was the answer. However, that interest quickly waned as more and more of us found out just how much of an investment it took to make those precious few contacts. Many of us found that the cost per QSO, whether monetary or time, was way beyond our means. This form of propagation has all but died out for the time being on this band."*

In the three decades since the first 50MHz amateur moonbounce contact between the teams of Dick Allen, W5SXD, and Joe Muscanere, WA5HNK (now W5HNK) in Texas, and Connie Marshall, K5WVX (now K5CM) and Sam Whitley, W5WAX (now K5SW) in Oklahoma, the playing field has changed dramatically. That historic contact utilised a pair of 8-Yagi arrays, each aimed at the horizon.

Previously only an option open to a very few of the world's largest stations, moonbounce contacts are now within reach for most well-equipped 6m stations. Until recently, an accomplishment such as making a moonbounce contact on 6m EME was about an S-unit or two beyond the reach of most stations. And when it comes to something like 6m moonbounce, where signals are just on the threshold of detectability, 5 to 10dB is a *huge* amount! However, several very significant developments have combined to make this out-of-this-world type of 6m propagation practical today.

6m is indeed the most challenging band in the VHF / UHF spectrum due to the physically large size of 6m antennas, the high background sky noise, the many geomagnetic disturbances that can deflect the moon-bound and / or earth-bound signals off course, and the prevalence of local noise around 50MHz. But what has changed since 2000 is the arrival of WSJT. By taking advantage of recent advances in digital software, utilising the extra ground gain available to an antenna when the moon is near the horizon, and carefully scheduling EME operations around optimum times of the month, successful contacts can be made by modest-sized stations.

Operating EME is relatively straightforward using the same type of equipment and software as other VHF and UHF bands. The ready availability of transceivers and amplifiers that operate on 6m, along with improvements in antenna designs and coaxial cables, bring 6m EME capability within the grasp of anyone interested in trying it, albeit it will require rather more effort than working the other propagation methods. So, rather than complaining when there is no ionospheric propagation on 6m, why not aim your antenna at the moon and see what is possible?

The following paragraphs cover the basics. For further information, there is plenty of material on the Internet. W7GJ's [12] website is a great starting point. For those with an archive of *QST* material material or members' access to the ARRL website, the June 2005 issue also carried a primer on using JT65 for EME [13].

While it is quite possible to make EME contacts on 432MHz on CW using a relatively modest station, CW is only feasible on 6m with exceedingly large EME setups at both ends of the contact (a few such stations do exist). What WSJT has done is to bring 6m EME to the point where it is no longer the preserve of those with lots of real estate and / or deep pockets.

The logistics

The mean distance between the earth and the moon is 385,000km. The moon's diameter is 3476km so a little trigonometry shows it appears a mere 0.52° wide as viewed from earth. Bearing in mind the three-dimensional polar diagrams of typical amateur antenna arrays, it is obvious that only a very small amount of ERP will illuminate the disk. Moreover the moon is a sphere, not a flat mirror, so only radio waves that hit the middle region will be reflected back to earth.

In other words, the path loss is enormous. To put some figures on it, the minimum round-trip path loss at 144MHz is 251.5dB rising to 270.5dB at 1296MHz. On 6m the overall path loss will be around 242dB, somewhat less than on those higher bands, but still considerable. But in practice the figure is somewhat academic as there as so many other factors at work, with 6m being affected much more by ionospheric factors than are the higher bands.

Getting going on 6m EME

There are three areas that deserve consideration in order to ensure successful 6m EME contacts: equipment, planning and operation.

Failure to assemble a station that can effectively receive and transmit enough dB will result in lack of success. By the same token, if you can exceed the threshold required for success, you will suddenly be rewarded with the capability to make world-wide contacts on 6m at any time of year or during any point in the sunspot cycle.

One of the big challenges with 6m EME is the high level of urban noise if you live in a built-up area. This can make EME QSOs difficult with lower gain antennas, so living in the countryside is a very definite benefit! Of course, the main way of minimising this problem is to add elevation to your antennas, though this is no mean challenge with the large high-gain antennas required to undertake 6m EME.

Equipment considerations have been covered earlier in the chapter as they are, to a large extent, common to all WSJT operations (albeit that EME requires as much power as you can muster. But the antenna system requirements for EME are somewhat more demanding than for any other 6m activity.

First you must decide what type of antenna you will use. It is essential that a modern, computer-optimised design of Yagi is selected. There can be many dB difference between specific antennas, and on EME (especially on 6m EME), every single dB is critical. It is a waste of time to try to use one or two old Yagis just because they happen to be available behind the garage. Such a convenient antenna may provide acceptable results on strong Es or F2 propagation, but EME is a much more critical and far less forgiving judge! The typical single Yagi successfully used on 6m EME for horizon-only contacts is a long boom computer-optimised 7-element Yagi. One of the main UK antenna manufacturers, InnovAntennas [14], has many highly-optimised 6m antennas of all sizes to suit 6m EME needs.

If you are going to limit your activity to an hour or two around moonrise and moonset, and your antenna is going to be mounted close to the ground (less than 25m high) you probably will be much more successful by selecting the largest

Chris Gare's, G3WOS, 6m moonbounce antennas: two 7-element InnovAntennas LFA2 Yagis on 9.5m booms.

single Yagi possible. This will give you the lowest possible angle of radiation (essential for assuring 'common moon windows' with distant stations on the opposite side of the earth), and will provide the best utilisation of the much-needed 'ground gain' (up to 6dB) that is afforded to a Yagi aimed at the horizon. Ground gain is an essential element in the overall system performance of a modest EME station. When beaming at the horizon (moonrise and moonset) reflections from the ground in front of your antenna will add to the direct signal and give another 3 – 6dB or thereabouts, at certain elevations and depending on the local terrain, the ground conductivity and the height of the antenna above ground. Ground gain can make all the difference to achieving a solid contact. So the more modest stations don't try to track the moon but, instead, wait for the moon to move through the main lobe of their antenna. (Note that, unfortunately, ground gain is considerably reduced when you live in a typical UK urban environment with houses in all directions).

If you are planning to track the moon for more flexibility in scheduling other stations, you will need to assemble an array to make up for the lack of ground gain afforded by the single Yagi aimed at the horizon. You will also need some way to point the antenna upwards and to read the elevation. In the simplest arrangement, this is provided by ropes and protractor with a weighted string mounted to its centre. In general terms, it requires an array of four Yagis when elevated to match the gain of a single Yagi aimed on the horizon. This may be strictly correct from an antenna gain perspective, but elevation can reduce local noise by 8 to 12dB, vastly increasing the ability to decode JT65 signals. This is especially the case in typical UK urban environments.

Regardless of the type of antenna(s) you are using, it is essential that you use the lowest loss coaxial feedline possible, and that it be as short as possible. A good goal is to have the loss at 50MHz be under 1dB. Such goals are usually quite achievable with the new coaxial cables that are currently available (such as 4-50 Heliax) or by using 'hardline' for most of the run. If you are planning to use the cable in an application which will require repeated flexing (such as around a rotator), make sure you use one with stranded centre conductor (such as the LMR Ultra Flex) that is designed for such an application. The feedline data in Chapter 3 should help in selecting a suitable feedline for your location.

Planning

Because of the numerous variables involving different radios, computers and interfaces for them, feedlines and interconnecting cables, the complete station should be tested thoroughly before it is transferred to its final location or any attempt made at 'real' QSOs.

The specific days of the month that you choose to be operational can have a very significant bearing on your success. The EME signal 'degradation factor' changes continually as the moon progresses in its orbit around the earth each month. This factor varies depending on the distance to the moon and the amount of 50MHz radio noise being generated in the region of space behind the moon. The degradation factor is expressed in terms of additional dB loss, compared with the ideal conditions of the moon being at perigee (its closest point to earth), with a quiet background sky.

If you are constrained to a certain day because you are arranging a schedule with another station who also has a horizon-only antenna system, you will be at the mercy of the degradation on that particular day. However, if you are planning a DXpedition or have a steerable antenna, you have the ability to choose your operation times so they are best suited to days with the least degradation.

GM4JJJ's *MoonSked* software [15] is invaluable in planning the best times for EME communication. The *Moon Graph* output (see **Fig 6.3**) plots key factors including earth-moon distance (the smaller the better – the distance varies over time as the moon's orbit is not an exact circle), *TSky* (the noise temperature of the 'sky',

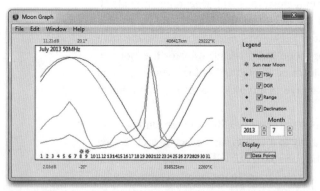

Fig 6.3: *Moon Graph,* from GM4JJJ's *MoonSked* software, shows the earth-moon distance, the sky noise temperature, degradation etc, plotted for each day of the month.

which varies throughout the year and will affect signal absorption), and degradation (already mentioned, a calculation taking various effects into account). In an ideal scenario, all would be optimum on the same date or dates – in practice this rarely happens!

6m EME signals can also be affected easily by geomagnetic disturbances. Even if the MUF is not high enough to produce ionospheric communication paths at 50MHz, ionisation can cause the signals to be deflected off their course. This can happen as they are heading out through the ionosphere on the way to the moon and / or as the incoming signals are returning to earth. Many times, it is not possible to predict when such ionisation might cause QSB or even complete signal blockage. However, in some cases (because the sun rotates every 27 days), it is possible to predict potential auroral disturbances that can affect high-latitude stations. Similarly, during the spring and autumn, it is possible to predict the positions of high ionisation that can cause TEP propagation and can upset EME propagation if one or both stations are located in the lower latitudes (the 'TEP zone'). Intense Es openings can also block the EME signals, and the typical azimuthal directions of such ionospheric aberrations often can be approximately predicted depending on the time of day and time of year.

Certainly, if the local noise is high at certain times of day, that can be a serious consideration for a receiving station. For example, operating in the evenings, when all your neighbours' TVs and computers are turned on, may be problematic and early morning skeds are often the best for minimum noise and least chance of TVI.

Another example of a local factor would be the 'trapping' of transmitted signals by tropospheric ducting or a local temperature inversion. The effects of such atmospheric effects are most noticeable on signals when they are transmitted at low angles along the horizon, and are minimised if the signals can be transmitted at higher angles.

Successful 6m EME operations are often characterised by those that have the ability to provide real-time or near real-time coordination. If both stations have access to the Internet, it is easy to decide on a clear frequency and confirm that both stations are operational and can be available to start a schedule. This is often more difficult with DXpeditions, which often depend on e-mail or satellite phones to pass on information regarding intended frequencies, transmitting sequences, and dates / times they will be operational. Because EME stations at one end of the path will often find that their chance to contact the other station is during the middle of the night, it is less compelling to gear up to try for a contact without any assurance that the other station will in fact be operational on a certain frequency.

Operating considerations
Especially on DXpeditions, special RF filters on audio leads as well as scheduling and frequency coordination with other operators may be necessary to avoid QRM from other nearby transmitters.

An elevation control is necessary to work EME when the moon is above the horizon. Another view of the G3WOS 6m EME array.

The location of the antenna should be as close as possible to the operating position to reduce feedline loss. If the antenna is to be a horizon-only operation, special consideration must also be given to locating the antenna where it will have the most ground gain. There should be clear, unobstructed terrain in the direction of moonrise and / or moonset to maximise clean reflections coming up from the ground toward the antenna. If a clear view over a lake or the ocean can possibly be obtained, it is definitely worth the effort to locate the antenna to take advantage of that.

The standard mode for 6m EME operation is the WSJT mode of JT65A. It is very helpful to practice with the software and become experienced making contacts before the first EME operation. Practice sessions can be arranged with other local VHF stations or on HF. HF practice partners can be found at the JT65 terrestrial chat room [5].

It is recommended that you use the established EME procedures and standard messages, so as not to confuse your QSO partners or make it more difficult for them to copy you.

W7GJ adds this advice: "In the way that the standard messages are structured in WSJT, the contact is completed by sending the standard message following the one that you receive. For example, if you receive the first message (callsigns), you select the second message (callsigns and OOO reports) to be sent. If you receive the second standard message of callsigns and reports, you send the third message (RO), which indicates receipt of the callsigns (by sending the O report) and receipt of the report (by sending R). Especially if you have a limited moon window (caused by one or both stations being restricted to horizon-only operation), you cannot afford to make the mistake of transmitting back the same message that you have just received; the contact progresses by *alternating* messages between the two stations.

"It also is essential that both stations transmit during each of their scheduled sequences throughout the entire schedule period. Unlike ionospheric propagation, where it is often best practice to wait until you can copy the other station before you call them, on EME it is *very unusual* for both stations to be able to copy each other at the same time. This is because Faraday rotation changes the polarity of the signals going out through the earth's magnetic field to the moon and back. Therefore, one station may be copying fine while signals are cross polarised at the other station, attenuating them by more than 20dB! That is the main reason schedules are run on EME – so that the required information can be sequentially exchanged between stations to eventually result in a contact."

QSL for the first EME contact between the late SM7BAE and ZL3NW (see the sidebar over the page).

SIX METRE EME, A ZL PERSPECTIVE
(by Rod Mackintosh, ZL3NW)

Six Metre EME was something I had not considered until Kjell, SM7BAE, e-mailed me in 2000. He knew I had recently put up a 10-element Yagi on a 13.2m boom at the top of my 25m tower and he considered a CW EME contact may be possible. Back in 1988 ZL2BGJ had worked WA4NJP on 6m EME but he used a wire rhombic antenna and I had not considered that a single Yagi would work. At that stage Kjell had an array of four Yagis for Six. We ran many skeds and often we could receive part contacts, but the first of several completed contacts was on 5 January 2001.

In about 2003 Joe Taylor, K1JT, produced several computer programs, one of which is suitable for weak signal work and in particular EME communications. The mode JT65A has been widely used throughout the world and has made 6m EME contacts possible with relatively small stations at both ends, i.e. single Yagis with good ground gain.

The amount of ground gain, which may be up to 6dB, is dependent on foreground ground conductivity. The maximum antenna gain lobe is typically at a small elevation to the horizon and this is dependent upon the height of the Yagi above the ground. With a single Yagi there will typically be minor antenna lobes as well. For example my 10-element Yagi at a height of 25m has a main antenna lobe at about 3°. However there are several other minor antenna lobes up to about 18.5°. When the moon is at 3° I get maximum signal from the moon but I have worked big stations up to 18.5°. It is fortunate that often the moon is at low elevation in New Zealand and Europe at the same time. That is, moonset in New Zealand is often moonrise in Europe. It should be appreciated that a single Yagi with ground gain and no elevation is similar gain-wise to four similar stacked Yagis with elevation where there is no ground gain.

A good example of ground gain is demonstrated by working David, MM0AMW, with a single Yagi. There are both moonrise and moonset widows between David's QTH and mine. At David's moonset he is looking over the Atlantic Ocean and at his moonrise he is looking over land. That is, there are two EME windows to work New Zealand and while his moonrise window is typically longer it is far more difficult to complete an EME QSO. I have worked David both ways but his moonset over the ocean offers far better signals and I believe this is due to the superior reflection over the ocean. I have also had several great 6m EME contacts with Jan, OY3JE, with his single Yagi for similar reasons.

I have had many successful EME contacts after heavy rain or when there is snow on the ground and I believe this to due to better ground conductivity and hence better ground gain.

The other requirement is low noise and often stations in city areas have problems with this, although I have worked several stations in cities, so it is possible. Often it will take several attempts and agreeing on a clear frequency to be used.

I have had many memorable 6m EME contacts but one was completing a 6m EME contact with Ian M0BCG, now G5WQ. This was the first G - ZL 6m contact at a distance of 19,002km and occurred on 24 January 2005 at 1607UTC. Ian runs 4 x 5-element Yagis with elevation. I have had many contacts with Ian since the initial contact and it is appreciated when Ian often helps with monitoring and reports after my attempts with others smaller stations.

I have, however, noted the variation with which two stations quite close to each other may receive me. There are lots of variables with 6m EME, but one major one is Faraday rotation of signals and this can cause major problems should the signals be rotated close to 90°.

The greatest distance I have worked was with Joel, F6FHP, who had a single 9-element Yagi. The distance between us is 19,441km. At this distance, which is near the antipodes, the moon was at less than 4° during the common window, which was of just a few minutes duration.

Faraday rotation and one-way propagation

This is not the place to go into huge technical detail about EME operation, as most readers of this book will probably never attempt it, and those who do will find help and resource material on the web and from dedicated 6m EME players. But it is worth being aware, in simple terms, that there are a lot of factors which can affect a signal on the long journey to the moon and back, some of which have already been mentioned.

In addition, the polarisation of the signal can be affected such that instances of apparent 'one-way propagation' occur, where the station at one end is able to decode the far end's signal OK, but the reverse is not the case. On the EME path, polarisation can change quickly, in a matter of minutes, which can disrupt what was on course to be a perfectly good contact. The subject is covered on, for example, the W7GJ [10] and G3WOS [16] websites, and the most serious of 6m EME operators have installed antenna systems where they can switch polarisations as required or, since the introduction of MAP65 as part of the WSJT software suite, take signals from two orthogonally-polarised antennas and combine them using MAP65 to overcome Faraday rotation effects.

But it's a big challenge. It's bad enough having to put up a two- or four-Yagi array for EME, without having to double it up in terms of polarisation! As a result, although MAP65 is now being used regularly for 2m EME, at the time of writing no one is using it on 6m, but serious 6m DXers are nothing if not ambitious, so this may well change in time. For a fuller explanation of Faraday rotation refer, for example, to *Wikipedia* where it is covered in some detail [17].

Results

Don't expect to work the world on 6m EME, at least not immediately! However, there are a few of very well-equipped four-antenna 6m EME stations who are active regularly and prepared to make skeds. Most of these folk have designed their stations with the aim of being able to work a 400-watt station with single Yagi at the far end.

Occasionally expeditions to rare locations will make an effort on 6m EME to make a particular DXCC entity available on 6m when terrestrial propagation is unlikely. 5W0GJ, T32C and VK9/ZL1RS are three examples of expeditions in recent years that have done just that. Generally these expeditions will have relatively modest 6m EME set-ups, so if you aspire to work such expeditions you will need to aim for something like a

Lance, W7GJ, activated Samoa on 6m EME only, in August 2011.

The four-sided QSL from T32C, a mainly HF DXpedition that also operated on 6m EME.

multi-Yagi antenna at your end, along with as much power as your licence allows (some countries, including the UK, make higher-power permits available, for example, under certain circumstances and on application).

Fig 6.4 is a screenshot of a 2009 EME QSO between K1JT and RU1AA using JT65.

Finally, check out *Wikipedia* [11] for more about EME, and take a look at the websites of IW5DHN [18], N1BUG [19] and DF9CY [20], all of which can be recommended for those contemplating EME operation.

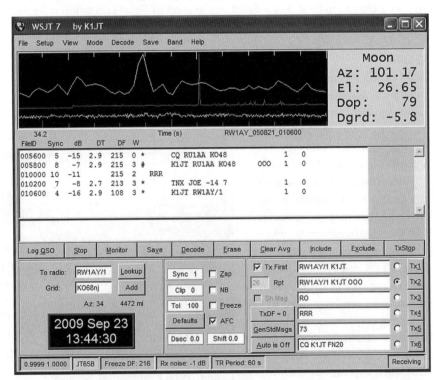

Fig 6.4: A JT65 screen shot of a 2009 EME QSO between K1JT and RU1AA.

USING WSJT DURING Es AND F2 OPENINGS

Six metre operators are increasingly turning to WSJT during marginal openings via multiple-hop Es and F2 propagation. The principles are the same as for meteor scatter operation. Contacts will usually be pre-arranged (even if only by a matter of moments) by way of the Internet, ON4KST, or Ping Jockey, agreeing frequencies and timings. A look at the All-time table on the JT6M webpage shows the sort of distances that have been worked by this means, for example W5UWB to IS0GQX via multi-hop Es in June 2008 at a distance of 9414km.

The following advice on using JT6M during such openings is from the JT6M webpage, with thanks to Kev Piper, G0CHE.

Trans-Atlantic Openings

If you are fortunate to be active during a multi-hop Es opening across the Atlantic or to other DX destinations, or a TEP opening, it may be worth while giving JT6M a try. Distances of over 8000km have been worked in this way.

The following information is provided to help all JT6M users interested in making DX contacts and should be taken as guidelines only - adherence to these guidelines will provide everyone with the greatest chance of good DX QSOs. These guidelines have come about through discussions between G4IGO, G4PCI, WA5UFH, and his WSJT group including K1SIX, N5SIX and W5UWB.

Frequency

It is worth noting the differences in calling frequencies between NA (50.260MHz) and EU (50.230MHz). There needs to be some collaboration as to which frequency to use. For Europe, 50.230 may cause a problem with QRM to those not in the DX path but within those who are. The same could also apply to NA with 50.260. During Es openings in Europe 50.230 sometimes gets smothered with SSB QSOs, making this frequency unworkable. Please remember, 50.230 and 50.260MHz are designated as calling, not operating frequencies.

CQing

In Europe the norm is to call CQ with an operating frequency, e.g. "CQ 238" – meaning the caller will listen on 50.238 for any calls and will QSY there to complete the contact when a call is heard. In NA the norm is to use U or D + number to indicate QSY frequency, e.g. U5 or D10 implying they will listen up 5kHz or down 10kHz and QSY to transmit there when a call is heard. It is important to understand the two different methods so that successful QSOs can be completed. As mutual QRM is inevitable if the band is open, please use the QSY frequency all the time when calling on the calling frequencies.

Transmit periods

Typically the western-most station will transmit during the first 30-second period. This means that for NA and Caribbean openings Europe will always transmit second period. Conversely openings to the east (from Europe) should imply the reverse, although at the time of writing this has yet to be experienced on a large-scale opening.

Reporting

It has been suggested that everyone uses the European method of reporting (this is based on IARU Region 1 MS procedure) . As Es propagation provides strong signals at times and mostly for the duration of each over, it is also perfectly acceptable to use the standard method of reporting rather than usual MS procedure type, e.g.55 or 57 etc.

The important thing is to receive both calls and exchange reports (26 through to 59) or your grid square for a valid contact. This is entirely up to the operator, but it may be wise to follow your QSO partner's method if they send a report first. (i.e. if you are still sending calls only).

Propagation

With long-distance Es (and F2 in future years) propagation, signals will be subject to QSB – fast, slow, deep etc. This suits JT6M very well, and it may appear to be like ionospheric or tropospheric propagation at times with signals present but very weak: these conditions normally give a good decode. At other times the signal strength may rise rapidly and be S5 or more (maybe time to grab the key or microphone?) – either way, if the path is there a contact should result. You may experience deep QSB whereby you lose the signal from your QSO partner. Be patient, the signal may well come back again to allow the QSO to continue and complete.

* * * * *

While the above paragraphs refer specifically to JT6M, the new mode JT9 is also being used for the same purpose – weak signal terrestrial band openings (see the following section).

OTHER DATA MODES

In closing this chapter, it is worth a short review of a couple of other datamodes which are often encountered on 6m. All interconnect with the radio and computer in the same way as described earlier in this chapter in the context of the WSJT software and, in the same way, are 100% duty-cycle modes during transmit, so beware that you adhere to the advice of your transceiver manufacturer in terms of how much power you can safely run. While some are not, strictly speaking, weak signal modes, they have much in common in that the PC is doing much of the work.

WSJT-X JT9

JT9 is a component of the WSJT-X software suite, an experimental program release deriving from WSJT. At the time of writing it is still in its infancy but a number of people are starting to make QSOs on 6m using JT9. The *WSJT-X User Guide* (downloadable from K1JT's website [1] says, "Version 1.0 of WSJT-X offers a new mode called JT9, designed for use on the LF, MF, and HF bands. JT9 shares many characteristics with the modes JT65 and JT4 made popular in WSJT. All three modes are designed for making reliable, confirmed QSOs under extreme weak-signal conditions. They use nearly identical message structure and source encoding. JT65 was designed for EME ('moonbounce') on the VHF / UHF bands and has also proved very effective for world-wide QRP communication at HF; JT4 is used mainly on the microwave bands. In contrast, JT9 is optimized for HF and lower frequencies. JT9 is about 2dB more sensitive than JT65A while using less than 10% of the bandwidth. World-wide QSOs are possible with power levels around 1W and compromise antennas. Several dozen JT9 signals fit easily into a 1kHz slice of spectrum." Although the preceding extract makes it clear that JT9 was developed specifically with HF operation in mind, it is already (May 2013) being used quite extensively on 6m to make terrestrial contacts under what might otherwise be marginal propagation conditions.

PSK & BPSK

While by no means a weak-signal mode in the same way as the various WSJT modes, PSK (and BPSK) is easier to use for 'ragchew-style' contacts and yet can be very effective at low power levels or when signals are weak (but, generally speaking they do need to be audible; don't expect them to recover signals that are well below the noise level in the same way as WSJT). PSK is used quite extensively on both 6m and 4m as a search for datamodes spots on *DX Summit*, for example, will reveal.

BPSK stands for Binary Phase Shift Keying and the most popular amateur system is BPSK31 (usually referred to simply as PSK31) [21]. It was developed by Peter Martinez, G3PLX, specifically as a replacement for traditional RTTY for real-time contacts. Peter's argument was that modes using error correction were inherently unsuited for real-time contacts, not for technical reasons, but because the users tended to be uncomfortable with the time delays inherent in hand-shaking between the two ends of the contact. The '31' in the name refers to the number of characters sent per second. The character set is based on a number of 'varicodes', i.e. variable-length codes, devised in a similar way to Morse code, whereby the more common letters can be sent more quickly than the less common ones.

Without going into any more detail, the benefits to the user of PSK31 compared with traditional RTTY are that it tends to work better in weak signal conditions and uses less bandwidth (a single carrier, as against the two of RTTY, with a theoretical bandwidth of 31Hz).

PSK31 requires a stable transmitter and receiver, but this is not a problem with modern transceivers. In any case, most of the software available for PSK31 uses a waterfall display and incorporates DSP filtering. Therefore operating the mode is simple. You can leave your transceiver's filters on a wide (SSB) setting and watch something like 2.5kHz of the band. This is room aplenty for many PSK31 signals. You can then click on an individual signal to decode it and start a QSO. Unlike RTTY, it doesn't matter whether you transmit and receive in USB or LSB mode, as there is only one tone, so there are no issues about tones being reversed. And if you invoke automatic frequency control (AFC) it will stay locked to the signal, even if there is some subsequent drifting in frequency.

PSK63 [22] is a variant of PSK31 operating, as the name suggests, at a higher speed. It has been developed very much with contesting in mind, with the intention that contacts can be conducted that much more quickly. As you would expect, it takes up roughly twice the bandwidth of PSK31. Most of the popular software programs available for PSK31 also support PSK63. Programs include *Digipan, WinPSK, Winwarbler, Multipsk, PSK31 Deluxe, QuickPSK* and *RCKRTTY*. As well as text, PSK63 is capable of sending thumbnail pictures, with a transmission time around two minutes.

Other modes

While there are many other datamodes (and new ones appearing all the time), the only other mode encountered with any regularity on 6m and 4m is traditional RTTY. While something of an anachronism nowadays, RTTY enjoys huge popularity on the HF bands as it still lends itself better than most datamodes to quick-fire two-way exchanges. Perhaps it is because there are so many operators familiar with RTTY that it survives on 6m and 4m in the face of more recent and sophisticated datamode programs.

REFERENCES

[1] WSJT page: http://physics.princeton.edu/pulsar/K1JT

[2] 'I Don't Use Digital, Honestly I Don't', Ken Osborne, G4IGO, *SixNews*, Issue 94 (February 2008).

[3] JT6M: www.jt6m.org

[4] Dimension 4: www.thinkman.com/dimension4/download.htm

[5] JT65 Terrestrial Link (chat page): www.chris.org/cgi-bin/jt65talk

[6] ON4KST: www.on4kst.com/chat/start.php

[7] Ping Jockey: www.pingjockey.net/cgi-bin/pingtalk

[8] International Meteor Organisation: www.imo.net

[9] DL1DBC: www.dl1dbc.net/Meteorscatter

[10] OH5IY: www.kolumbus.fi/oh5iy

[11] Wikipedia on EME: http://en.wikipedia.org/wiki/EME_(communications)

[12] W7GJ: www.bigskyspaces.com/w7gj

[13] 'EME with JT65', Gene Zimmerman, W3ZZ, *QST* June 2005, pp80 – 82.

[14] InnovAntennas: http://innovantennas.com

[15] GM4JJJ MoonSked: www.gm4jjj.co.uk/MoonSked/moonsked.htm

[16] G3WOS (Faraday Rotation): www.gare.co.uk/6m_antenna

[17] Wikipedia on Faraday rotation: http://en.wikipedia.org/wiki/Faraday_rotation

[18] IW5DHN: www.qsl.net/iw5dhn/when.htm

[19] N1BUG: www.g1ogy.com/www.n1bug.net/prop/eme.html

[20] DF9CY: www.df9cy.de/ar/radio.htm

[21] PSK31 'Official' Homepage: http://bipt106.bi.ehu.es/psk31.html

[22] PSK63: www.qsl.net/kh6ty/psk63

7 A 6m persective from around the world

THIS CHAPTER LOOKS AT 6m from around the world. It does not cover 4m which, for the most part, is largely a European band for the time being. 4m propagation and expectations have been covered elsewhere within these pages. But it is worth saying that, as 4m availability spreads to more distant parts, much of what is said in this chapter about 6m will apply equally.

The 6m DXer has to have a number of psychological characteristics such as patience, persistence and a strong level of motivation. Exactly how these play out in practice depends on where in the world you live. If you live in one of the main centres of 6m activity (US, Japan, Western Europe) you can, should you so wish, rely nowadays on Internet-based sources for DX alerts. The various chat rooms and *Cluster* systems which are discussed elsewhere in this book can provide the necessary alerts to much of the DX that is around. But that doesn't mean that you will necessarily hear the DX or, if you do, that you will be able to work it. Propagation, as discussed elsewhere in these pages, can be very localised and your near neighbour may be able to hear a DX station that is completely inaudible with you. So even when you are alerted to an interesting DX station you may still have to sit on his frequency for quite some time before his signal emerges from the noise (or perhaps doesn't!) And if it does, because you live in an area of high 6m population, you may well be competing with lots of other stations to make a contact during what might turn out to be a very short opening. It can be quite a challenge.

In contrast, if you live somewhere like South America, Southern Africa or Oceania, the challenge can be very different. There probably won't be others in your area busy spotting the DX and you will have to hone your technique for finding it under your own steam. Typically this might mean programming your VHF receiver with a selection of commercial stations and beacons (VHF TV or radio carriers, for example) in the direction in which you are interested and scanning these continually so that you can immediately notice when the MUF starts to climb in that direction, or, as was mentioned in the chapter on equipment, using a modern SDR receiver to monitor key parts of the spectrum below 50MHz.

Some DXers have this down to a fine art. Take a look, for example, at the list of South African 'firsts' on the ZS6EZ website [1]. These are dominated by just three DXers although there are quite a few ZS amateurs active on Six. How did these top

ZS 6m DXers manage this? Primarily by knowing what DX was, or might be around, and carefully monitoring a range of frequencies in that direction to spot any potential propagation opportunities. A knowledge of propagation didn't hurt either, knowing, for example, which times of the year were best for TEP propagation. Part of the trick was also knowing which countries were about to gain access to the band, less of an issue nowadays as most countries have that access and many of the 'firsts' have already been made (although DXpeditions may still be the first to make a serious effort on the band from wherever it happens to be).. But one way to be first in the log back then was to call CQ during a band opening and work the pile-up in the hope that you would be called by that elusive 'new one'. After all, if you were one of the few ZS stations (for example) on the band, you would almost certainly be a 'new one' for him too.

Watching out for DXpeditions is obviously key, too, provided they have made provision for 6m operation. Given the narrow beamwidth of high-gain 6m Yagis, these expeditions may not be beaming to parts of the world where amateur activity is sparse. The good news is that many DXpeditions nowadays are available via the Internet, perhaps by e-mail or even in real-time on the ON4KST chat pages. So, with a bit of effort and coordination, a DXer away from the main population centres may still have a decent chance at working some new ones. And, of course, there is always EME as discussed elsewhere.

SIX FROM EUROPE

Let's start with Europe. This section will be short because much of the relevant information has been covered already in this book. In recent years almost all European countries have gained access to Six, even though there are still restrictions in certain countries in terms of frequency, power or who is allowed to oper-

San Marino is always a popular 6m catch within Europe, thanks to regular activity by Tony Ceccoli, T77C, who has achieved DXCC on Six.

ate. At one time or another every European country but one (Franz Josef Land) has been on the band and worked, although some, such as Mt Athos, are active only occasionally, and Russian stations are currently absent from the band.

European DXers also have a good path during their Es season to the northern countries in Africa (places like Morocco, the Canary Islands and Egypt), to much of the Middle East (Israel, Lebanon, Cyprus etc) and, in a remarkably reliable way, to the Caribbean and the northern part of South America. Openings to Japan appear to be a regular feature around the summer solstice (see Chapter 4), particularly from south-eastern Europe, but occasionally extending to western Europe. Stations located in southern Europe also benefit from regular TEP propagation, enabling them to work further down into Africa, the Indian Ocean and South America and later, over the long path, into the Pacific Ocean and Asia.

With such a head start it is not surprising that European amateurs have come to dominate the higher echelons of the DXCC 6m standings, with scores of more than 250 entities.

The Pacific countries are the toughest from Europe with openings only possible at the peak of the sunspot cycle. Again, propagation is dependent on location. Only two KH6 contacts are believed to have been made on 6m from the UK, whereas stations in Mediterranean locations such as Malta and Greece regularly (at the top of the cycle) experience long-path KH6 openings of several hours duration either in the morning or in the evening. Japan is worked by multiple-hop Sporadic E every year from southern and eastern Europe, whereas from western Europe the path is open much less frequently. In recent years, northern Europe has also experienced short Es openings to Alaska, KL7, over the North Pole.

SIX FROM CENTRAL ASIA
(with thanks to Nodir Tursoon-Zadeh, EY8MM)

In recent years the main activity from Central Asia has been by EY8MM, EY8CQ, EY7AF and UK9AA. For the record, their stations are as follows:

EY8MM:	Ant: 7-ele Yagi
	Rig: Icom IC-7800
EY7AF:	Ant: Vertical
	Rig: FT-1000MP MkV + FTV-1000
UK9AA:	Ant: 4-ele quad
	Equipment not known
EY8CQ:	Ant: 7-ele Yagi
	(Now QRT from EY, living in Russia).

There has also been activity by UN and EX stations. Nodir maintains a record of 'firsts' on his website [4] but says that the future of 6m in EY is uncertain as a Channel 1 TV transmitter affects those regularly active EY amateurs. Nodir himself reached 101 entities for his 6m DXCC during the 2006 Es season to achieve the first 6m DXCC from Central Asia. He was able to add a few more entities during 2007 and 2008 during periods when the TV transmitter was off the air for maintenance. The hope is that when the TV service transfers to digital, regular 6m activity can resume.

Nodir Tursoon-Zadeh,EY8MM, in his well-equipped shack in Dushanbe, Tajikistan (MM48jn). EY8MM is one of the very few stations in Central Asia active on 6m.

SIX FROM JAPAN
(with thanks particularly to Han Higasa, JE1BMJ)

Japanese amateurs were granted the use of 6m in 1952, although there is evidence of activity in the 1937 – 38 period on 56MHz by stations in the J2 (Tokyo – Yokohama) area. It is possible that after WWII the first 6m QSOs were made by occupying Allied forces personnel, as Japanese nationals were prohibited from amateur radio activity until the end of the occupation in 1952.

Six is extremely popular in Japan and, for many amateurs, is their entry band to the hobby. There are probably more than 2000 JA amateurs active solely on Six, with about a quarter of these from the JA1 (Tokyo) call area. Many other JA stations operate on Six from time to time, especially now that so many HF radios have 6m capability. But because of space and other limitations there are probably no more than about 500 JA amateurs with the capability to work true 'weak signal' DX on Six. Some JA 'firsts' are shown in **Table 7.1.**

Around 1967 SSB started to become popular on 6m, helped along in 1968 by the introduction of the first dedicated 6m transverter, the FTV-650, from Yaesu. Before then almost all QSOs had been made on AM or CW. That same year the first VHF SSB meeting took place in Tokyo, and members decided to adopt the calling frequencies of 50.075 and 50.575MHz.

For many years the power limit at VHF was 50 watts, with no antenna limitations. Since 1999 the limit on 50MHz has been 500 watts, but a 1kW licence is available under certain circumstances.

Unlike Europe, there are not too many DXCC entities within one hop of Japan. Those that are include HL, DU, JD1(Ogasawara), JD1(Minami Torishima), BV, BY, UA0, JT, XX9, VR2, KH2 and KH0.

TEP openings are usable from JA7 and the south of Japan, but rarely from

May 1953	JA1EE	– JA1DI	110km QSO on 50MHz
Jul 1953	JA1FC	– JA6BV	First Es QSO on 50MHz
Apr 1954	JA8AJ	– JA3CQ	1000km Es QSO
Jan 1956	JA1AHS	– VK4NG	First DX QSO
Mar 1956	JA6FR	– LU9MA	First SA QSO, 19,000km
Oct 1956	JA1AUH	– K6EDY	First NA QSO
Apr 1957	JA6FR	– PY2AXX	First PY QSO
Feb 1958	JA1AXE	– ZL1ADP	First ZL QSO
Mar 1958	JA4HM	– VK9BW	First New Guinea QSO
May 1958	JA1AXE	– ZS1SW	First ZS QSO
Oct 1958	JA3CE	– CT3AE	First long-path over 25,000km
Jan 2004	JE1BMJ	– WA4NJP	First 6m EME from Japan

Table 7.1: JA 'firsts'.

JA8. When TEP occurs (usually around the spring and autumn equinoxes), a typical opening might include:

Afternoon TEP   (03 – 06UTC, 12 – 15 local)   VK1 VK2 VK4 VK5
Afternoon TEP   (but limited to December / January) ZL
Evening TEP   (08 – 12UTC, 17 – 21 local)   VK4 VK6 VK8 P29 YB

Notwithstanding these limitations, the leading JA in the 6m Honor Roll listings is JA1BK with 223 credited. There are others close to or above the 200 mark, but this is largely the result of good propagation during previous sunspot peaks, when openings to Europe, Africa etc were common. F2 paths at sunspot peak include:

- North and South Pacific (short path)
- North America (short path and skewed path)
- South America (short path)
- Asia (short path)
- Middle East (short path and skewed path)
- Africa (short and long paths, depending on location and conditions)
- Europe (same as Africa, skewed path)
- Oceania (TEP, short path, skewed path)

Skewed paths are almost certainly the result of scattering from high electron density regions around the equator. Okinawa, JR6, being south of the main Japanese islands, generally experiences better propagation that the rest of Japan, with more frequent openings.

Perhaps one of the most interesting propagation phenomena is the JA to Europe path around the summer solstice, which seems to be reliable even at sunspot minima. Some argue that this is the result of multi-hop Sporadic E, but Han, JE1BMJ, believes that multiple hops over land rather than seawater would result in such high attenuation that the path would be unworkable. Instead he has proposed what he describes as SSSP (Short-path Summer Solstice Propagation) which he describes in an article that appeared in the UKSMG's *SixNews* and has already been mentioned in Chapter 4.

JI1CQA's website [2] has a list of 6m 'firsts' from Japan and shows that 261 entities had been worked as of May 2013. A detailed history of 6m in Japan appears in English on JA1RJU's website [3], translated from Japanese by Roy Waite, W9PQN.

A new band plan for 6m came into being on 31 March 2009. Japanese hams operating narrow bandwidth data modes such as WSJT, JT65A, JT44x and RTTY are restricted to operate between 50.300 to 51.000MHz. There are two exceptions:

when they are replying to DX (non-JA) stations they can use 50.0 to 50.1MHz; when on EME they can use 50.0 to 50.3MHz. DXers elsewhere in the world looking for JA contacts should be aware of these restrictions.

JE1BMJ says, "*I have heard many JAs are working with VKs in JT65A around 50.090 MHz; many JA–JA QSOs are made on 50.330MHz – this is the de facto main channel for JT65A or JT65HF activity in Japan*".

SIX FROM NORTH AMERICA
(with thanks to the late Gene Zimmerman, W3ZZ, and kindly updated by Chris Patterson, W3CMP)

The USA has had a 6m allocation from 50 to 54MHz since amateur radio resumed after World War II in 1945. Six metres is a popular operating spot for thousands of US amateurs and is aptly called the 'Magic Band' because of its eclectic propagation. The relatively recent advent of HF / 6m transceivers has increased the level of activity here. In recent VHF contests upwards of 4000 or more operators have participated on 6m. A significant number of these have stations and antennas capable of working long distances on 6m under marginal conditions.

The late Gene Zimmerman, W3ZZ, in typical pose.

The power limit on 6m is the same as it is on most other bands: 1500 watts output with no licence restrictions on antennas. Towers are generally limited to 61m height by aviation law without special permission and conditions. However, local zoning laws and Covenants, Conditions and Restrictions – 'CC&R' – may limit erection of towers and antennas. US amateurs use all forms of transmissions both analogue (SSB and CW) and digital (most usually forms of WSJT like FSK441, JT6M and JT65a). A small number of stations have legal limit power and huge antenna systems with up to four 15m boom Yagis either stacked vertically or arrayed on an H-frame. Such large arrays include the multi Yagi systems at K1TOL (FN44), K2MUB (FN21), K3TKJ (FM28), contest arrays at K1WHS (FN43) and K8GP (FM18). These arrays can be used on EME and can work single Yagi stations with booms no more than 9m in length, particularly when both stations use JT65a digital.

KH6/K6MIO claims the longest contact at 19,360km and N6CA at 18,464km, the longest in the contiguous 48 states.

The US is geographically large and diverse. Only a relatively small number of countries in the Caribbean, Central America and off the Canadian East Coast are reachable within the 2000km E-skip or the 5000km F2 single hop radius. But, because of its size, essentially all types of VHF propagation can be seen at one time or another in many places throughout the country. Long distance contacts are possible via 'E-skip' (Sporadic E) propagation around the solstices and via F2 when the sunspot numbers reach levels high enough to generate solar flux

Chris Patterson, W3CMP, on a 6m DXpedition in the Caribbean.

indices over 200 for a few days at a time.

The more recently discovered propagation mode, SSSP (Summer Solstice Short Path), has enabled contacts between Japan, China, and the eastern US in late June and July; the corresponding Southern Hemisphere mode has allowed contacts between Australia, New Zealand and other South Pacific entities and the US.

Transequatorial propagation at about the time of the equinoxes is possible via spread-F to South America directly for locations at about 30°N or below and via Es links to TEP for areas north of that. Auroral propagation is common during periods of high geomagnetic activity in the more northerly latitudes especially above 40°N but can extend southwards to 30°N at K indices of 8 or 9. Meteor scatter propagation over distances of 2000km is readily achievable everywhere and even tropospheric ducting is known, though it is much less frequent than on frequencies of 144MHz and above.

Propagation on 6m resembles that on 10m but is much more erratic because the MUF is twice as high. At the peak of the sunspot cycle much, but not all, of the world is workable from almost all of the US via F2. Within the US F2 propagation is relatively long and tends to bridge the east and west coasts both laterally and diagonally. During the summer months much of the Northern Hemisphere is available to some part of the US. In general, Sporadic E is more common the farther south you go. Almost the entire contiguous US is reachable by two Es hops. During the equinoxes, South America is readily workable by TEP or Es-TEP links.

Some parts of the world, notably in central and South-East Asia, are very difficult to work from any part of the US because of propagation and perhaps because of lack of activity or licensing. Specifically among these are Zones 17 – 19 (Asiatic Russia); Zone 22 (VU and 4S7); Zone 23 (JT and UA0Y); Zone 24 (BY, BV, VR2 etc), and Zone 26 (South-East Asia). The East Coast, particularly W1 – 3 and northern W4s, have problems working Japan and other northern Far Eastern countries. It is just in recent years that SSSP has been utilised to make these contacts. The West Coast and particularly the Pacific Northwest have problems working Europe because the path is directly through the auroral zone. The Midwest also has problems to Europe because of the auroral zone. The Caribbean and northern South America are easy to work from the East Coast and the South, more difficult from the Midwest and not easy from the West Coast especially via Es because all contacts are multiple hop. The same is true for Europe. However, the path to Europe from Alaska is extraordinarily difficult and has been bridged only twice via any mode of propagation; until the past few years the Pacific Northwest has not been able to work Europe via Es and that has now happened but only a few times.

PACIFIC NORTHWEST PERSPECTIVE

(As the previous section makes clear, North America is a large continent and there are very different experiences of 6m, especially for those in the northwest. The following, with thanks to Lance, W7GJ, gives a very particular perspective from somewhere that is on the fringe of most 6m propagation)

While European openings from the Midwest (W8, 9, 0) are an annual occurrence during the Es season, life is much tougher from the Pacific Northwest (geomagnetic latitude over 54° N). Lance reports (July 2008), *"We did enjoy several good Es openings to the south-east this summer, and we did actually get enough Es clouds properly lined up for some multi-hop openings into the Caribbean. I even was able to contact Curacao and Venezuela on multi-hop Es this summer, along with FJ, PJ6 and KP4. Many of us up here in the Pacific Northwest tried constantly, day after day, for some sign of life from CY0X to no avail - it is extremely rare for Es clouds to all line up along the Canadian border so as to provide 'hard bounces' to cover the 4300km path to that island"*.

What Lance had not enjoyed, due to his location in the mountains, which is blocked by a 13 degree horizon toward EU, was a terrestrial opening to Europe. This changed in the summer of 2012 when a multi-hop Es

A view of the W7GJ VHF antena farm: on the left 16x17-element Yagis for 2m, in the centre a 6M11JKV 11-element 6m Yagi on a 70ft long boom and 70ft high, and on the right 4x6M9KHW 9-element Yagis for 6m.

Lance, W7GJ, in his 6m EME control centre.

opening lasting over four hours allowed him to work: G0JHC (at 1345UTC), F9IE (1352UTC), DK3WG (1402UTC), DL3BQA (1419UTC), SP3AGE (1421UTC), SP7AWG (1423UTC), SP3OCC (1424UTC), SP4MPB (1432UTC) and SP3RNZ (1759UTC). Lance emphasises that this was not a first for Montana or the other states involved in the multi-hop Es opening in 2012, but was a first for him due to his particular location. He does point out, though, that for such an opening the Kp index needs to be low, and it has been remarkably low most of the time – even during the 'peak' of the recent sunspot cycle. Lance says, "*I am confident that we will see more of this type of propagation in the future, with the excellent coordination available through the ON4KST 6m real time chat pages on the Internet*".

Lance's experience is that the geographic location on the earth and the geomagnetic latitude play a large role in determining the likelihood of Es or F2 openings from a particular location. From his location there are rarely links down to the TEP zone. Neither is there such a thing as 'scatter path' or 'skew path' or such terminology associated with more southern forms of propagation. Everything from that far north is direct and almost always requires what Lance refers to as a 'hard hop' – a true MUF of 50MHz in the middle of the path. This is in comparison to the Caribbean or east coast USA, which enjoy frequent chordal propagation (which is possible at much reduced MUF levels) along ionisation contour lines toward Europe. These contour lines are clearly shown on the Space Weather website and correlate quite well with propagation at times of day when they are lined up along great circle paths. It is worth noting that the TEP propagation usually happens in the spring and autumn, and that Es in the northern states happens primarily in late June and also rarely around Christmas. That is why links from this area to TEP are very rare. But, says Lance, "*on March 29, 2012, we did have a rare Es opening to the south in the afternoon. As the ionisation drifted off the West Coast, a link formed to the TEP / F2 zone that provided contacts to FK and VK9N for number of 6m stations here in Montana, Idaho, Washington and Oregon. That, in fact, was the only ionospheric DX I worked to the South Pacific during this sunspot cycle*".

From the Pacific Northwest, 6m operators enjoy propagation along such contour lines only in the direction of Japan, which occurs late in the day local time. However, it is possible for them, like the rest of the USA, to participate in evening chordal propagation to Japan if there happens to be an Es cloud to the north-west to force the chordally-propagated signals down to the earth. It is very rare to have

Es north of their already far north latitude, but it is quite common for stations in other parts of the USA to have Es to their north-west and, as a result, to contact Japan on this chordal Es propagation with good signals.

On 4 May 2013 the SFI rose over 145 and Lance believes this probably moved the TEP further north than usual. With the early onset of Es, which happened to connect him to Texas just at the right time of late afternoon, he made his first ionospheric links to South America on TEP this solar cycle. On 4 May he completed CW contacts with CX3AN, LW3EX, LU2DPW, LU5FZ, CX9AU, LU5FF and PY3FF, while CX1FK, LU2DEK, LU9DO, CX5CR, LU6DC, PY3OR and LU9EHF were worked on SSB. On 5 May the SFI was still over 135 and he worked CE2AWW weakly on CW. He is still searching for HC, CP and PZ.

Lance concludes his interesting overview by saying, "*I hope the above information provides a picture of the 'Magic Band' from the Pacific Northwest. As you now understand, it is a long way to other DXCC entities from here. Only Canada and Mexico can be contacted on single hop Es. All other entities require at least double hop Es or F2 and, even then, there are not very many countries within reach! But 6m has become my favourite band, and it is very exciting because of the different types of propagation that are possible. In the 12 years that I have been on the Magic Band, I have contacted a total of 184 different DXCC entities, including 129 via EME*".

SIX FROM SOUTH AFRICA
(with thanks to Chris Burger, ZS6EZ)

South African amateurs have had the use of 6m since the late 1940s and were able to work into Europe during that period when several European countries, albeit briefly, had 6m access (as chronicled elsewhere in this book). The list of 'firsts' on ZS6EZ's website [1] shows G and VE in 1947, for example, while the first JA - ZS QSO was recorded in 1958. But from then on until the late 1980s, when many more countries started to gain access to the band, South African amateurs had a fairly quiet time on Six, with even their near neighbours such as 3DA0, 7P and A2 being absent.

Sporadic E is notable in ZS largely by its absence, so there isn't the annual Es excitement that JA, European and North American amateurs enjoy. 6m activity is also concentrated in a relatively small area (about 1500km across), requiring high levels of ionisation before contacts become possible. Couple these constraints with the low number of active stations, and it is easy to understand why Es is not an everyday occurrence.

One of South Africa's top 6m operators, Chris Burger, ZS6EZ.

The latitude of ZS means that TEP is quite reliable and has been studied quite extensively, albeit more at 144MHz than at 50MHz. But the huge distance from ZS to the major centres of amateur radio population has meant either waiting for F-layer propagation to come along or to struggle during the sunspot minimum by working near-in DX via meteor scatter and the more distant DX by EME. Even square-chasing is harder than in Europe or North America, not only because of the lack of Es, but because the squares are larger! This is because the nature of the QTH locator system is such that squares occupy a constant unit of longitude, but that gets narrower in absolute terms the nearer one gets to the poles (the effect is lost on a Mercator map projection). However, starting in the mid-80s, there has been considerable square-bashing by a number of amateurs, resulting in firsts from most of the neighbouring countries. Most of these expeditions conducted meteor scatter schedules using SSB, serving a limited audience of perhaps a dozen DXers.

The ZS6EZ website lists 153 'firsts' (included four 'deleted') from ZS and it is notable that 116 of these were made by just three 6m enthusiasts: ZS6WB, ZS6AXT and ZS6NK. Two lessons emerge – ZS6 is further north than the ZS1 area where the other major centres of ZS ham population lie. And to catch the rare ones on Six you have to be well set up for the band and dedicated to being there at the right time. Only a few hardy souls tend to qualify, whether through personality or circumstances (e.g. being able to work from home). Catching those new ones from ZS would have required an up-to-date knowledge of which countries were likely to appear on the band, an eye on propagation (for example, by monitoring for European commercial stations between 30 and 50MHz to spot rises in the MUF) and a decent station to catch the opportunities as and when they occurred.

In the *6 Metre Handbook*, published in 2008, the following comment was included: "Now that 6m is well-established world-wide, ZS amateurs are well placed to work some new ones when the sunspots return, but in the meantime face a fairly quiet time on the band". In practice, as a result of the poor solar peak, just four new countries have been worked by terrestrial means since then – T6 in 2012 (one solitary contact), 4O and A9 in 2011, plus HA in 2006 (initially via EME, worked later by TEP). As a result, most of the die-hards are now active mainly on EME, using digital modes, especially WSJT, as the only way to catch new ones until solar activity rises sufficiently.

SIX FROM BRAZIL
(with thanks to Peter Sprengel, PP5XX / PY5CC)

PY is an ideal location for 6m operating, in a very good position relative to the geomagnetic equator. Don, PY5ZBU (ex-G3MWM), is believed to be the first person in the world to have confirmed 100 entities on 6m but his cards went astray *en route* to ARRL and Don lost interest in the band after that. So the first from South America was PY5CC who, as mentioned elsewhere, was at 200 countries by October 2000 (he currently has 221 credited).

The very successful November 2012 PT0S DXpedition from St Peter & St Paul Rocks. Top 6m DXer Peter Sprengel, PP5XX / PY5CC, was a member of the four-man team.

A list of PY 'firsts' appears on the 50MHz.com website [5] and shows 249 entities worked from Brazil as of May 2013.

Propagation from Brazil is largely via TEP, but because of the country's location, it is one of the first to enjoy F2 propagation when there is sufficient solar activity.

Amateurs elsewhere in the world are also in debt to the PY 6m guys for activating the three island groups that count separately from Brazil, i.e. PY0S (St Peter & St Paul Rocks), PY0T (Trindade) and PY0F (Fernando de Noronha). Some 134 entities have been worked from PY0F (the easiest to get to and operate from), 88 from PY0T and 39 from PY0S (by far the hardest to activate as it consists of just a few rocks, but this total benefitted from the excellent PT0S expedition in 2012). Peter made something like 4000 6m QSOs from PY0FF's QTH during several visits over a period of about 10 years.

There are only a handful of Brazilian amateurs active on 6m, but enough to ensure that Brazil is very workable from both Europe and North America on a regular basis.

SIX FROM AUSTRALIA

(compiled with help from Steve Gregory, VK3OT / VK3SIX, and the UKSMG archives)

6m in Australia dates back to the immediate post-war period, though there was a short period when the band changed to 56 – 60MHz. During the 1958 International Geophysical Year, a 50 – 54MHz allocation was reinstated, but from 1964 to 1989 this was reduced to 52 – 54MHz to avoid conflict with TV channel 13.

Callsign	Date	Worked	Country	Distance
VK5KL	26/08/1947	W7ACS/KH6	Hawaii	9000km
VK2WH	30/12/1953	VR2CB	Fiji	4000km
VK4NG	22/01/1956	JA1AHS	Japan	8000km
VK4ZBF	05/02/1958	K6RNQ	USA	13,000km
VK4NG	20/03/1958	KX6AF	Marshall Is	6000km
VK4NG	25/04/1958	LU8OL	Argentina	12,000km
VK6ZAV	19/04/1958	VS2DO	Malaya	6000km
VK6CL	06/08/1958	DU1SW	Philippines	5500km
VK2ADE	07/04/1959	VE7AQQ	Canada	13,000km
VK3ALZ	02/05/1959	XE1FU	Mexico	13,400km

Table 7.2: Significant Australian 'firsts'.

Some significant firsts, listed in an article in the UKSMG archive, are shown in **Table 7.2.** The final contact shown in Table 7.2 set an Australian distance record which was to stand until 3 April 1979, nearly 20 years later, when it was increased by a mere 300 kilometres.

Australia is a big country, so the experience of 6m varies from one call area to another. But VK amateurs enjoy all the propagation types that have been discussed elsewhere in this book, including aurora (but via the 'southern lights' in their case). Stations in VK6 and VK8 are much more likely to enjoy F2 QSOs with Europe compared with those in VK2, VK3 and VK5. In contrast, the further north and east the VK station is located, the better the chances of working the USA.

What VKs do enjoy, along with the ZLs (and making European amateurs green with envy) are openings into the Pacific, to all those little islands (though, in fairness, 6m activity from some of those places is few and far between). But it's

Australian and New Zealand amateurs enjoy openings into the Pacific on 6m – when there is activity. The 3D2R (Rotuma) DXpedition was active on 6m in September 2011. Colin Bradley, WA2YUN, is resident on Wake Island, and runs a personal 6m beacon.

tough from VK and during sunspot minima EME can be the only way to catch new ones. Steve, VK3OT / VK3SIX, is the highest-placed VK on the DXCC listings, with 131 countries confirmed, a total that a well-equipped European 6m operator could achieve in just two or three seasons. In Steve's own words, *"It always amuses me to see European operators comparing their country tallies. It's easy when you have 100 countries within range. I wish they would stop asking us how many countries we have worked this year!"*

SIX FROM NEW ZEALAND
(contributed by Rod Mackintosh, ZL3NW)
Although there is recorded activity from 1934 within New Zealand on the then 5-metre band (60 MHz) it was not until after WWII that activity increased on 6 metres (50 – 54MHz).

There was wide availability of surplus valves such as the QV04/7, 832a, 815, 829b and 807. The 807 was used for both RF and modulators. Modes then were CW and AM. Surplus equipment was also in abundance, for example RF26 units fed into R1155A, AR88 and ARC-5 receivers. Antennas were simple dipoles up to a maximum of 4-element Yagis.

It is documented that ZL4GY created a world record by working the USA in 1949. The USA callsign is not given but no doubt the station was on the west coast. Shortly after this, contacts were made from New Zealand to stations around the Pacific Rim. In 1953 Graham, ZL3GS, and Bob, ZL3NE, worked

Some Pacific DX worked on Six by Rod, ZL3NW, over the years.

NEW ZEALAND 6-METRE MILESTONES

1934: First two-way aeronautical mobile made on 5 metres (60MHz) between ZL4BV and ZL4CA.

1935: TV experiments authorised on the 5-metre band and on wavelengths below 1 metre.

1939: WWII closed down amateur radio for the duration.

1945: 8 December at 7.30pm amateur radio resumed. 58.5 – 60MHz allocated.

1947: First two-way contact between the North and South Islands.

1947: Distance record of 635 miles made by ZL4BK and ZL1HY.

1949: World record contact on 50MHz between ZL4GY and USA (no USA callsign given).

1953: ZL3GS and ZL3NE work VR2CG (Fiji).

1953: 26 December ZL3GS worked VK9DB (Papua, now P29). VK9DB reports this was his first ZL contact on 6m.

1958: 13 February ZL1ADP worked JA2IF for first Japan to New Zealand QSO.

1959: At midnight on 31 January 1959, 50 – 51MHz withdrawn for channel 1 TV.

1979: 50 – 50.15MHz sharing arrangement introduced.

1980: 16 November ZL3NE worked VE1AUX at 15,555km.

1982: Non-Morse licence operation permitted on 51 – 53MHz.

1988: 8 October world record 6m CW EME contact between ZL2BGJ and WA4NJP at 13,256km.

1990: New arrangements and special conditions announced for access to 50 – 51 and 53 – 54MHz.

2001: 5 January 6m CW EME world record extended to 18,014km (SM7BAE and ZL3NW).

2001: 3 April terrestrial world record between ZL3VTV/1 and EH7KW: 19,921km (SSB used).

2005: 24 January first 6m contact between England and New Zealand using digital (JT65A) EME, M0BCG and ZL3NW, at 19,009km, a world EME record.

2006: 3 March 6m digital (JT65A) EME record extended to 19,441km (F6FHP and ZL3NW).

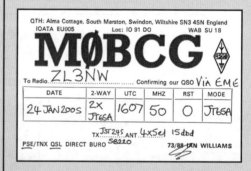

QSL from M0BCG (now G5WQ) confirming the first UK to ZL QSO on 6m, made by EME on 24 January 2005.

Wynn, ZL3DX, who was operating as VR2CG from Fiji (now 3D2). Also that year Graham worked Doug, VK9DB, in Papua Territory. This was Doug's first ZL contact.

Looking at the log of Fred, ZL1ADP, it is interesting to see the level of activity when he was first active in 1957; there were clearly a lot more active 6 metre stations than today. In 1958 Fred worked his first JA, thought to be the first JA -

ZL contact, and he also worked his first USA station, K6RNQ. Most areas of Australia were worked at this time too.

The introduction of television and more particularly the channel 1 frequency allocation of 44 – 51MHz (video 45.25, sound 50.75) in 1959 limited the use of the 6m band in areas where channel 1 was, and still is, in use. Initially 6m could be used in these areas outside TV broadcast times. Later television was extended to 24 hours per day, and the introduction of permits limited the number of operators who may operate in the bottom megahertz of the band. These permits have only been issued to amateurs located outside the channel 1 TV area, though it should be noted that 51 – 53MHz may be used throughout the country. For some years activity was around 52MHz, with the calling frequency on 52.050, but nowadays the bottom 200kHz is widely used internationally and hence almost all ZL activity is also there. Currently there are about 15 operators within New Zealand who are active on 6m, spread from the far north to Dunedin in the south. There are large areas in the country with no activity due to channel 1 TV. However, it should be noted New Zealand channel 1 TV is a great propagation indicator for DX stations!

Although the Pacific Rim was worked as far back as 1949 to 1958, it was not until 1980 that this distance was extended, when Bob, ZL3NE, worked VE1AUX at a distance of 15,555km. On 3 April 2001 at 2003UTC Tim, ZL3VTV/1, made SSB contact with Jose, EH7KW, at a distance of 19,921km. Signals peaked at a direction of 145 - 150° and the signal strength peaked at S3.

A very interesting contact was also made on the same day when Raj, VU2ZAP, was worked long path by Ray, ZL2KT, and Morrie, ZL2AAA, at 2002UTC. Bob, ZL4AAA, in the far north also copied Raj and worked him the next day at 1947UTC at a beam heading of 160°. There was no signal on the direct path of 290 - 300° and Raj confirmed Bob's signal peaked via the long path of 315 - 320°. Raj was last heard at 2015UTC and peaked at S9.

I will mention a few interesting contacts I have made. The first was with Duarte, CT3HF, on 28 April 2000 at 2046UTC. I remember it well as I was letting the dog out and fortunately I had my remote 'baby monitor' with me. Out of the speaker came Duarte calling CQ. I rushed back to the shack and by then John, ZL3AAU, was working Duarte. I worked him next followed by Ross, ZL3ADT. Duarte then faded away but it did show the value of a remote monitor and how short in duration some long-distance contacts can be.

The contacts to northern Italy and some of Europe on 3 February 2002 will always be remembered by the ZL3s in the Christchurch area. Prior to this it was often wondered if such contacts were even possible. Signals peaked between 0930 and 1000UTC. Initially I could hear Bob, ZL3TY, on the west coast working DX but the ZL3s on the east coast could not hear what he was working. As time went on I started to hear some of the DX stations. Later signals peaked to very good SSB copy and even above my severe power line noise. Duncan, ZL3JT, with his excellent CW and operating skills worked several countries during this opening.

The contact with Phil, FJ5DX, on 6 April 2002 at 2005UTC was an interesting one too. I was beaming LP to Europe at about 170° when I heard Phil calling "CQ beaming to Africa" and I went back to him. We had time to check that there was no direct path and oddly enough although there were other locals on I was the only one to hear and work Phil.

It was great to work Nick, YA4F, on 2 and 4 December 2002. His signal was 539 for both contacts. On these two days there was no sign of any VK6 beacons on the same beam heading but on the 3rd they were quite good signals but with no sign of Nick. On the 4th I heard Nick calling CQ for some time and it even gave me time to telephone John, ZL3AAU, some 24km away to see if he could hear him but there was no trace at his QTH. As Nick kept calling I decided to try another contact as he had no takers. One positive thing about my second contact was that it gave Bob, ZL3TY, on the west coast a chance to find him on my frequency and thus he worked him after me.

During the better sunspot cycle peak years it is not unusual to have propagation follow the sun around the Pacific Rim. First thing in the day would be to look for any indicators of long path to Europe at about 150 to 170°. After this and a little later in the morning turn the beam to about 90° and listen for the XE1KK beacon. This beacon is an excellent propagation indicator and is often heard before an opening to the Caribbean, California or an extended opening to Florida. Early afternoon look for any JA beacons, although as often as not the appearance of JA stations is the best indication that the band is open. Late afternoon, keep an ear open for any Hawaiian beacons. It is interesting that although the direction to Hawaii is similar to California the propagation times are quite different, i.e. there

The 6m long-boom Yagi at the ZL3NW station.

are different propagation modes involved. During the mid-afternoon there are often openings to VK1, 2, 3, 4, 5 and 7. Late afternoon or early evening an opening to VK6 and 8 and if one was very lucky openings to Europe mid to late evenings at about 270°.

During the sunspot minima years there is still some propagation via Sporadic E during December and January and to a lesser degree November and February. Contacts may be made to all of Australia and any South Pacific Islands which are active. The VK channel 0 TV is a great early indicator. For example the video on 46.240MHz is often received first, followed by the sound on 51.740MHz. This is then often followed by the VK2RHV and VK2RSY beacons. At the peak of the Sporadic E season I have heard the VK6RSX beacon at Dampier for hours on end at a distance of some 5717km. Unfortunately there are currently no active 6m stations in Dampier, but it would be interesting to know how far beyond the west coast of Australia propagation extends. About this same time John, VK6JJ, worked Paul, A35RK, at a distance of 6971km.

Winter Sporadic E typically offers propagation during June and July and there are openings to the east coast of Australia, VK1, 2, 3 and 7 with the occasional opening to VK4 and 5.

It is interesting to note that although there has been 6 metre activity in New Zealand for many years, to date no-one has worked 100 countries.

Beacons:			
Call	**Freq**	**Location**	**Grid**
ZL3SIX	50.040	Near Christchurch	RE66ej
ZL2SIX	52.490	Near Blenheim	RE68vl
ZL2WHO	50.024	Waipuna Ridge	RF70om

The ZL3SIX beacon is outside the TV channel 1 area, hence the frequency allocation, and is the most widely heard ZL beacon. It should be noted there are other allocations for 6m beacons but they are not active.

The big change in the past few years is the phasing out of the analogue TV and in particular ZL TV Channel 1. The phase out is due to be complete on 1 December 2013. The ZL TV video transmitters with all their various frequency offsets have been heard far afield and have served as beacons for possible 6m openings. For example Bob, K6QXY, often listens for them and with the various frequency offsets around ZL he can tell what part of ZL the band may be open on 6m. Once they get up to about S3 at Bob's QTH he typically finds the MUF has risen up to 6m and I have had several contacts when Bob has given a 'heads up'. Likewise Bob, ZL1RS, in the far north has had many contacts this way when different frequency offsets are heard.

The closing down of the ZL analogue transmitters is a mixed blessing. They are lost as beacons but it means that once they are closed down all of ZL will be able to use 6m down to 50MHz. It is therefore expected there will be many more 6m stations active within ZL.

TV PROPAGATION INDICATORS IN ZL

TV Channel	Frequency (Nominal)	(MHz) (Actual)	Audio Freq	Location	Grid square	EIRP kW
NZ Ch 1	45.240	45.239600	50.740	Hamilton	RF72ul	250
NZ Ch 1	45.250	45.249983	50.750	Wellington	RE78js	200
NZ Ch 1	45.250	45.249992	50.750	Invercargill	RE43hx	300
Aus Ch 0	46.172	46.171684	51.672	Toowoomba	QG53tc	490
Aus Ch 0	46.240	46.240008	51.740	Wagga Wagga	QF35we	325

NEW ZEALAND SOUTH ISLAND PERSPECTIVE
(from Bob McQuarrie, ZL3TY)

As far as I know the following were the first EU contacts from the South Island:
9 November 2001: SP4MPB at 0730UTC on CW, SP6GWB at 0745UTC on CW.
Then on 13 November 2001: IK2GSO at 0839UTC on CW, I2YSB at 0844UTC on
CW, F1IXQ was heard but no QSO.

All this was preceded by the first ZL – 5B QSO when I worked 5B4AGM on 23
February 2001. Subsequently I worked 5B4FL on 31 December 2001 at 0712UTC
and JY5NX at 0801UTC. In this case the band stayed open until after 0915UTC but
no other QSOs were made. Then of course there was the opening of 3 February
2002, when I worked 19 European stations in I, YU and 9A.

I have never worked long path; the nearest was on 28 February 2002 when I
heard EH7KW and he heard me but we were unable to complete a QSO.

SIX FROM SOUTH-EAST ASIA
(with thanks to John Plenderleith, 9M6XRO)

John, 9M6XRO, has been active on 6m since mid-2005, using a Yaesu FT-920,
originally at 100W and, from 2010 onwards, with a Yaesu Quadra VL-1000 amplifier
producing 400W. His antenna is a 3-element quad on the roof of his apartment
block, 82ft (25m) above ground. He usually monitors 50.110MHz when he is in the
shack.

John's location is in Kota Kinabalu, the capital of the Malaysian state of
Sabah, on the island of Borneo (locator OJ85aw), and less than 6° north of the
equator. Being so close to the equator means that 6m DX openings correlate
closely with the summer solstice in either hemisphere so, as a general rule, Europe
is worked between May and July, and Australia etc from November to January. At
other times of year John says he frequently sees spots of JA–VK QSOs on the *DX
Cluster*, but cannot hear *either* side of the QSO: the signals skip over Borneo.

John has now worked 52 DXCC entities on Six, no mean feat when the closest real activity is from Japan, at a distance of between 4000 and 5000km! John has now worked over 2000 JA stations on CW, SSB and RTTY (a good proportion of those active on Six from Japan – see page 139). Elsewhere in the region John has worked stations in Ogasawara, South Korea, the Philippines, West Malaysia, China, Taiwan, Hong Kong, Indonesia, Vietnam, Mongolia, Uzbekistan and Kazakhstan. The 9M0L (Spratly), XX9E (Macau), and recently 9N7SZ (Nepal) DXpeditions gave him new ones, as did V84SMD (Brunei) and 9V1TT (Singapore), both of which had to obtain special permits to operate on Six. Other Middle East and Asian DX stations worked include 4Z4TL, A92IO and EY8MM.

However, it took John until June 2008 before he worked into Europe – that was with UT7IL. Other notable stations worked that year were E77DX and 5B4FL. In 2009 EA6SA was a new one. Good DX openings into Europe are few and far between, though. One such was on 24 June 2011, when he worked 9A2SB, ER0FEO, HA7TM, LZ1QI, OE4VIE, S57TW, SP8AWL, UK8OM, YO3BL and YU1AU – all new DXCC entities for John at the

The **9M6XRO antenna: it is a 'Lightning Bolt' HF bands 2-element quad, modified with three additional elements for 6m.**

time – plus other southern Europeans. An even better opening occurred a few days later, on 1 July 2011, when John contacted 9H1BT, A45XR, CT1HZE, EA7HG, OK7XX, T77C as well as VU2RBI – again all new countries for him. John's best DX on Six came towards the end of that opening, when he worked CT3HF in Madeira at a distance of 13,444km. John commented, "In this region DU1GM and DU7/PA0HIP are extremely active on Six but the openings we encounter are often quite different. I listened with frustration one time as they worked into Scandinavia and I could hear nothing of the DX. A few days later the roles were reversed as I filled my boots working into Southern Europe and finally CT3HF in Madeira with George and Willem on the sidelines not copying the DX. That's the 'Magic Band' for you…"

During that same opening, John contacted DL1YM, which is as close as he has come to working into the United Kingdom, though he was informed later that his signal was heard in England that day, peaking at S2 for a short time.

2011 and 2012 turned out to be the 'best' years on Six so far, with an average of 500 QSOs on the band in both years. The first four months of 2013 have generally been disappointing, though a notable opening was to KG6DX, KG6JDX and KH7Y. During that opening John also heard the personal beacons of E51WL and WA2YUN/KH9, but unfortunately both were on 'auto' and the operators were not around. Other stations worked in the Oceania area include VK9XS, FK8CP and 5W1SA, plus all the VK call areas. John reports that around December VK8MS is often S9+ on CW and SSB.

On 2 May 2013 9M6XRO worked SP5CFD at around his sunset. What made this a very peculiar contact was that the band was otherwise 'dead', with no beacons audible and no telltale 'vids' buzzing. Six is indeed the Magic Band!

REFERENCES

[1] http://zs6ez.org.za/lists/50zs1st.txt

[2] JI1CQA: http://ji1cqa.sakura.ne.jp/index.shtml

[3] JA1RJU (history, in English): www5.big.or.jp/~ja1rju/history_japan.html

[4] EY8MM: www.ey8mm.com

[5] www.50mhz.com/py.html

8 Portable and expedition operating

IF YOUR HOME location isn't ideal for 6m or 4m, one option is to head out to a suitable hilltop location and operate portable. You may do it simply for fun and as a change, to participate in one of the various portable contests that take place on the VHF bands, or maybe to try to catch some rare DX that you would be unlikely to work from home. The good news, for those of us in the higher latitudes, is that the main Sporadic E season occurs during our summer, when the weather is (hopefully!) more conducive to portable operation.

Even if you don't intend to operate portable or head off on a DXpedition, this chapter ought to be of interest in giving some insights into what is required to mount such operations which you, from home, will enjoy chasing and working. Indeed, by stopping to think about what is involved from the 'DX end' of such an operation, you may well gain insights into how best to catch a DXpedition on the band.

Much of what has been written elsewhere in this book applies equally to portable operation, but there are other factors which you may need to take into account. It all depends on what you have in mind. Portable operation covers a multitude of scenarios, from backpacking to hilltops with a handheld radio to mounting a full-scale DXpedition to an inaccessible, inhospitable island.

Your choice of location for a portable operation will be governed by a number of factors. If you want to work an announced DXpedition, for example, you may head for a seaside location, as the take-off over the sea gives a real boost to the low-angle component of your signal (lower loss in the near field of the antenna means that the important low-angle signals suffer less attenuation than over normal ground). If you want to operate a contest when conditions are likely to be average or flat, you may want to head for a hilltop location that is central to the various high levels of activity (surrounding towns and cities). In other words, you will need to give some thought to the ideal location for what you have in mind and then, ideally, make some reconnaissance beforehand to check factors such as accessibility, the need for permits (if any), facilities for camping if that is part of your plan, and so on. Every case will be different. But there are a number of factors which are common to most portable operations and these are discussed below.

PERMITS AND LICENSING

As this book is intended for an international audience it isn't possible to give specific details of what permissions are going to be required for portable operation. Everywhere will be different. But, generally speaking, anywhere that is privately owned will require you to have the permission of the landowner before you start setting up antennas. In the case of DXpedition operations from rare locations, those permissions may well need to be in written form which will satisfy the various awards bodies that you were indeed where you claimed to be and with the necessary authorisations.

As far as licensing is concerned, if you are operating from your own DXCC entity there is usually no issue, though you may need to sign with some sort of portable designator (this is no longer mandatory in the UK, though it is required by the rules of some contests). There is a lot of material available about licensing in different countries (e.g. [1]), and the CEPT licensing arrangements have made it possible for many amateurs to operate from any one of over 100 DXCC entities that are now on the CEPT list. This is the good news. The bad news is that 6m, and especially 4m, are not necessarily included in such arrangements, and it is necessary to check before firing up on the band and perhaps incurring the ire of local amateurs who, understandably, don't want to see their licence conditions treated in a cavalier fashion.

An example of the kind of situation that has to be dealt with is the 3B7C expedition of 2007, of which the author was a member. The standard Mauritian licence (3B6, 3B7 and 3B9 are legally parts of Mauritius though counting separately for DXCC) does not include 6m and, indeed, the 3B7C licence, when it arrived, did not include the band. Some negotiations with the Mauritian licensing authorities ensued and it became apparent that the reason for the lack of 6m authorisation was a concern that any operations would affect VHF TV in adjoining nations (other countries in the Indian Ocean region). In the end a compromise was reached, whereby three discrete frequencies were allocated to the expedition. This was fine, as these could be publicised to the DX world beforehand and there was, of course, no restriction on listening frequencies.

DXPEDITIONING ON 6M AND 4M

Some hardy 6m enthusiasts have been mounting 6m expeditions for years. Jimmy Treybig, W6JKV, and Lance Collister, W7GJ, spring to mind as classic examples. Once the number of amateurs active on Six started to increase, with the release of the band in many European and other locations, a 6m expedition became an interesting proposition, especially if planned to coincide with the best time of the year for long-haul propagation, usually June / July to take advantage of the main Sporadic E season in the northern hemisphere, where the majority of 6m DXers are located.

Once 6m was added to the ARRL DXCC Challenge as a qualifying band the demand for 6m operation during DXpeditions became even greater, as all those serious HF DXers suddenly needed a 6m contact to keep their place at the top of

AH51ba   6m EME   NIUE

Niue would be rare on any band. This was the September 2012 6m EME DXpedition by Lance, W7GJ, worked in the UK by Kerry, G8VR.

the listings. As a result, many HF-orientated DXpeditions now include 6m as a matter of course. However, the decision process regarding Six is likely to be different to a dedicated 6m DXpedition for a number of reasons, such as timing – HF DXpeditions are often scheduled around the equinoxes for best low-band propagation, but this is not optimal for 6m except perhaps at sunspot maxima. The best window for 6m EME may be different again, as discussed in Chapter 6.

LOCATION, LOCATION, LOCATION

It is notable that when well-known 6m DXpeditioners head off to a rare spot, they often seem to make their own propagation. The local stations in that area (parts of the Caribbean, for example) may rarely be heard on the band and yet, when these visitors show up, they seem to have regular band openings. One of the main secrets to this is choice of location, though good antennas and plenty of power also help. The local amateurs will normally have their station at their home, which is probably in one of the local towns where they have their employment, where their children go to school, and so on. In other words, surprising as it may seem, amateur radio is not the main criterion for their choice of location! They may also be constrained in terms of power by EMC problems. This is where the visitor has the advantage. He, or they, can select a suitable mountaintop or beach location with a clear take-off to one or more of the main areas of interest (North America, Europe and Japan) and, hopefully, no near neighbours. Cliff-top sites are perfect. Not the sort of place you would want to build your house, but ideal for a temporary expedition station.

There is nothing to beat a site visit to find a good location, but there is a lot you can do beforehand if you are planning any sort of DXpedition. Examples include contacting others who have been to the area before, maybe e-mailing a local amateur in the vicinity of where you want to go and, nowadays, taking a look on *Google Earth* which often has some great aerial shots which can give you a very good idea of what to expect.

BE QRV

As has been mentioned frequently in this book, 6m openings can be fleeting and if you have gone to the trouble of setting up a portable or short-term expedition station for whatever reason, you don't want to miss any of those openings. But this is where the short-term operation should have the advantage. The local amateurs will have jobs and families to think about. Even when they become aware of a band opening they may have to get home from work and do the best they can from their less-than-optimal location. They may have some portable kit that they can throw into a car and drive to a hilltop, but by then the opening may well be over.

You, on the other hand, are there for the purpose of radio. You can be on or near the radio at an optimal location 24/7, at least for the duration of your expedition. Or, at least, this is what you should be planning to do, though there may well be distractions by way of sun, sand and entertainment.

The trick is to cater for this in some way so that the openings don't get missed. The usual solution is to set up and run a beacon transmitter for the duration of your expedition, so that DX chasers, many of whom may be looking out for you, will quickly become aware if the band opens. All they then need is some way to alert you to the opening, whether by texting your mobile phone, calling you on the beacon frequency between beacon transmissions or, in the case of a multiband expedition, calling you on HF to let you know. A classic example is the 9M0C Spratly Island operation of 1998 in which the author took part. We left a 6m beacon running 24 hours a day. One day the beacon frequency came to life when a loud Japanese station called in the middle of the day, a clear sign that the band was open. Kazu, JA1RJU, the expedition's 6m operator, was in the swimming pool at the time, but was quickly summoned by one of the HF operators who had heard the callers on 6m and Kazu arrived, dripping wet, ready to run the pile-up. Such is the dedication of the true 6m enthusiast!

COMMUNICATION

The previous section mentioned the importance of DXers being able to alert an expedition station to a band opening. Nowadays there are many ways of doing that, not only by amateur radio means but by way of the mobile phone system and the Internet. Expeditions to the more remote locations are increasingly taking satellite phones and, while not necessarily publicising the number to all and sundry, may, for example, nominate a friend back home to take calls and pass on any relevant alerts or feedback. There is also a lot to be said for having the *Cluster* network available at the expedition location and / or Internet facilities to connect to sites such as ON4KST chat, so that there is real-time feedback about what is

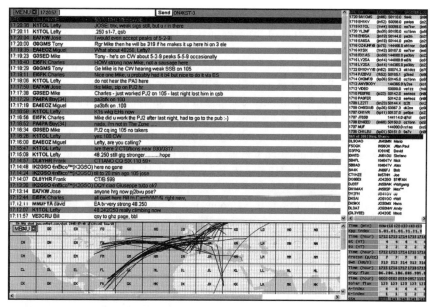

Example of the ON4KST 6m / 4m chat room .

happening on the band (to be able to watch propagation gradually moving towards you, for example). ON4KST is essential if 6m EME is included as part of the expedition plans.

WHAT FREQUENCIES & MODES TO CHOOSE

If your portable operation is for a contest, then choice of frequency and mode is a non-issue. You will simply go along with whatever is appropriate for the event. If, on the other hand, you are heading out on an expedition operation these are key considerations.

As a first example, suppose you are heading out to the Caribbean for a 6m expedition operation during the Sporadic E season. At times signals will, hopefully, be strong into the USA and possibly into parts of Europe. You will almost certainly be able to run stations faster on SSB than on CW and, at the same time, satisfy demand from those operators who are unable to operate CW. Unless your licence specifies spot frequencies you will need to choose one or more frequencies for your operation and, ideally, publicise them beforehand. But other expeditions will almost certainly be heading out at this key time of the year, so a degree of coordination is desirable. Certainly you cannot all expect to have access to the 50.110MHz calling frequency or chaos will ensue. So you have the option of trying to call CQ on that frequency and then announcing a move to another frequency or, as tends increasingly to be the case, announce a frequency before your departure and, unless there are good reasons to move (QRM on that frequency in certain target areas, clash with another operation etc), stick to it. This probably means using that frequency for all modes and expeditions are frequently to be heard on say, 50.103, running alternately on CW and SSB.

But signals won't be loud all the time. As has been mentioned elsewhere in this book, to make the most of Six you have to be prepared for true weak-signal working. The most common weak-signal mode is CW, which has a significant signal-to-noise advantage over SSB when copied through a good quality narrow bandwidth filter. So most expeditions will be geared up for CW, with suitable equipment and, more importantly, competent CW operators who can run through a CW pile-up at a fast pace for when those short, low signal-strength openings occur. A good example, during the 2008 Sporadic E season, was the TO5E St Barthelemy DXpedition which focussed on 6m (typically for such an operation, they had capability for the HF bands, to keep themselves amused when 6m was dead, but turned to 6m whenever there was the possibility of an opening). This activity made 1759 6m QSOs in 54 countries, a great effort, but 1435 of those contacts, a huge 82%, were on CW. In this case the operators chose to focus on CW, perhaps because of low signal strengths or maybe simply because they preferred that mode. Either way, as a DX chaser your best chance of a QSO was certainly on CW.

But what about those times when there is no F-layer or Sporadic E opening? Do you sit by the pool and relax, or do you try other modes of operation? This is a particular challenge for the HF expedition that also wants to operate Six, as the expedition may be taking place well away from the usual Sporadic E season. There are still options available, as for working from home, by way of meteor scatter and EME. Meteor scatter and other scatter modes can be useful if you are in scatter range of at least one of your major target areas. So MS would be a good option for a station in the Mediterranean wanting to work around Europe or a station in the Caribbean wanting to work the USA, as an example. But if you are in a more remote location EME may be the only viable option.

Setting up a 6m EME capability during an expedition is far from trivial. It's hard enough from a home location. But it has been done and to great effect, a notable example being the ZL8R Kermadec Island expedition in 2006, which was essentially an HF expedition from a very rare location, too far away from most major countries to rely on the more traditional propagation modes for 6m working. So the team chose to use EME and were able to make a number of QSOs by that mode, even into Europe, a very long haul indeed. The FT5XO expedition to Kerguelen Island also showed up on 6m and succeeded in working into the USA, a fantastic catch on the band. And, since then, as has been mentioned in a previous chapter, there have been several others, TX5K (Clipperton Island) being the most recent as this is being written in April 2013: a multiband operation where 6m EME specialist Lance Collister, W7GJ, went along specifically for that purpose. Lance has actually written an excellent primer on planning for a 6m EME expedition; it can found on (and downloaded from) his website [2].

PORTABLE OPERATING

How does operating technique when portable or on an expedition differ from that from your home location? Firstly, you are the station that others are chasing. So your time should be spent CQing, not waiting to hear signals. Most 6m DXpeditions, as mentioned earlier, set up a beacon at the earliest opportunity and leave it running when they are not making contacts. The usual system is that the radio goes back to receive during breaks in the beacon transmission and there should always be an operator nearby, even if he is reading a book and enjoying the sunshine, to hear any callers during those receive times.

At times when propagation is expected, the DXpedition will almost certainly be more proactive, calling CQ in the direction where propagation is most likely, and listening hard for the first sign of callers.

Once stations start to call the normal rules of DXpedition operating apply. If you (the expedition) have been calling CQ on the calling frequency of 110, now is the time to vacate that frequency and move elsewhere, probably to whatever frequency you have announced in your pre-expedition publicity. Once the pile-up starts to build, you will want to work split, probably with a

Keen VHF operator Bo Hansen, OZ2M, also operates on portable DXpeditions. Here is his 4m station on the Åland Islands, OH0/OZ2M, in May 2012.

reasonably wide split as you may be very weak with callers whereas they may well be very loud with each other, so you don't want their signals to be splattering across yours at the distant end. Similar considerations apply on 4m.

In terms of the QSO format, keep it short and sweet, for two reasons. Firstly, the opening may be very short and you want to put as many QSOs in the log as possible in the time available. If it's 'ragchews' you are after, then you shouldn't be on the 'sharp end' of a DXpedition! And, secondly, even within the length of a single QSO there may be significant fading, so keep the transmissions short. There is generally no need to include QTH locator, QSL information etc. This is all readily available via the *Cluster* network, your website and other sources. So just the callsign and the report. But if QSB does lead to a possible 'no QSO' it is worth hanging in there for a little longer in case the propagation takes an upward turn again and you can complete. Otherwise you could have some very disappointed customers.

In this respect, the better-equipped expeditions ensure they have an online log, updated regularly (at least once a day, so that the update is done before the next day's opening to the same area). That way DX chasers can see whether they are safely in the log and, if so, desist from calling again. But if they aren't, there is still (hopefully) an opportunity for them to try again. Bear in mind, though, that many DXers nowadays will be looking for a QSO on both SSB and CW. There is no fundamental harm in this provided that, in doing so, they and you are not depriving others from making their first contact with you. This is a judgement call only you can make, knowing how far you have already succeeded in working through the pile of potential callers from a given area.

Indeed, it is reasonable to ask how many 6m DXers there are around the world. There is no documented answer, but major DXpeditions at times when 6m is in good shape will typically make two or three thousand QSOs. The *Daily DX* website shows the current record for 6m to be held by VK9ML (Mellish Reef, 2002) with 4298 6m QSOs. It is not surprising that the record was set from that part of the world, as Japan has probably the highest number of 6m DXers (see Chapter 7). Over half that total was made on SSB, contrary to what I said earlier about the use of CW but a reflection of the fact that the DXpedition took place at the peak of the last sunspot cycle when there would have been some excellent, strong-signal 6m openings (VK9ML even recorded 146 RTTY and 56 AM QSOs on 6m during that expedition. Unfortunately their website does not show a breakdown by continent).

PUBLICITY

If you are heading off on a DXpedition you will want to publicise it in advance so that those who need you will be warned and can arrange to be around during your operation. Publicising a DXpedition has become dramatically easier in recent years thanks to the Internet. Generally all you will need to do is to e-mail the major bulletins and perhaps some specialist websites such as the UKSMG site. As far as bulletins go, the *Daily DX, 425 DX News* and *OPDX* bulletins are probably the most widely read and an e-mail to these three will ensure that news of your trip permeates to all the major areas. The important thing, though, is to send out that news in time for it to be picked up by national amateur radio magazines, which means allowing a good six to eight weeks before your trip to cater for publication deadlines. But at least, unlike in years gone by, you do not have to write to each magazine individually.

The other major element of pre-, during- and post-expedition publicity is to set up a website specifically dedicated to the operation. This is almost mandatory these days. A good website serves many purposes, advising the DX community of your plans (timings, frequencies, callsign etc), perhaps showing propagation forecasts for the main target areas, allowing log lookups once the expedition is under way, maybe offering a facility for requesting QSLs and / or making a donation and so on. Some of these facilities require that the website be updated once the expedition is underway, something that may be possible from your DX location, or may require a 'pilot' station back home who can work you on HF, perhaps, or via phone or e-mail, and then update the website with the latest news.

DURING THE EXPEDITION

Assuming you have some sort of Internet capability while on the DXpedition, it is increasingly common to put out news and upload logs on a regular (at least daily) basis. *Club Log* [3] is often the vehicle of choice for such logs, and offers the facility to include an OQRS (Online QSL Request) button.

Whether you choose to receive e-mails from all and sundry is a different matter – dealing with them can become very time-consuming and detract from the main expedition effort. On the other hand, it might keep you occupied between band openings!

QSLING, LOTW ETC

There is nothing unique about QSLing on 6m or 4m. As on other bands, those who work your DXpedition will want a QSL card. In designing the card, though, it is important that you include the QTH locator as this is something that matters to 6m DX chasers. There are no other major specific aspects to consider, but if your expedition was focused on 6m and / or 4m you may well want to include some information about the equipment, site, antenna(s), QSO statistics etc.

There are fewer 6m DXers and DXpeditions who use Logbook of The World (LoTW) compared with HF operators (and even fewer 4m operators). However, LoTW use is growing all the time and even though most 6m DXers will want your card for their collection, they may still prefer to claim as many DXCC credits as possible via LoTW in order not to have to risk their precious 6m confirmations to the vagaries of the postal service. So it is good practice to upload your expedition log to LoTW at the earliest opportunity. If you haven't used LoTW before, there are instructions on

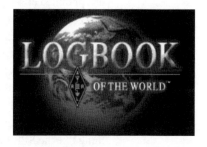

the ARRL website. But check before you head off as you may need to submit paperwork over and above the licence documentation to have your operation accredited, especially if you were operating from a particularly rare location or from somewhere that doesn't normally allow 6m operation. Again (see previous section) some expeditions have taken to uploading to LoTW while the expedition is still in progress. But the danger of offering to do this is that if, in the event, it becomes impossible, the world at large will inevitably become frustrated – always on an expedition, whether in this matter or others, set expectations which you have a realistic chance of meeting.

Whether you upload to other web-based confirmation systems such as eQSL is very much a matter of personal preference. eQSL does not have the in-built security systems that are integral to LoTW, but some DXers are happy to print a QSL from the eQSL database for their personal use, rather than have the expense of sending off for a more traditional QSL card.

EQUIPMENT CONSIDERATIONS

The chapter on equipment choice is a starting point but in a portable or expedition situation you may have restrictions in terms of the size or weight of equipment you are able to take, especially if you are flying to your destination. So while, for example, an FT-857 may not have the same RF performance as its bigger brother the FT-DX5000, it may be the more realistic radio for your expedition.

There is a particularly hard decision to make about whether or not to take a linear amplifier and, at the end of the day, this may prove impossible. But, given what has been said elsewhere about the low signal strengths on Six during marginal openings, a linear amplifier is undoubtedly an attractive 'extra' if there is any way that it can be taken along. It is no surprise that Chris Gare, G3WOS, has found the 6m DXpedition linear amplifier design on his website to be one of the most-visited pages [4]. The necessity of a linear amplifier was well illustrated in June 2008 when Dennis, K7BV, undertook a 6m DXpedition to Belize (V36M) and San Andres and Providencia (5J0M). His expedition amplifier went faulty shortly after his arrival in 5J0 and this undoubtedly had a major effect on the success of the DXpedition. For example, he worked very few UK stations, occasionally being heard at low strength. If the linear amplifier had still been operational, those openings would have been workable for significantly longer periods and at much better strength, allowing many more DX chasers to make it into the log. No doubt Dennis was very disappointed at what happened: with the best will in the world it's hard enough to carry one amplifier on a DXpedition and well-nigh impossible to take a back-up unit, so problems like this will inevitably occur from time to time. The good news is that, over the past few years, a number of lightweight amplifiers have become available from Ameritron, Elecraft, Expert and others which cover 6m at the 500 watt (or higher) level. Suitable amplifiers for EME are somewhat more of a challenge in that a power level of 800 watts or so is desirable (you will almost certainly be working with a single Yagi, albeit a substantial one) at a high duty cycle, so something like the SPE Expert 2K-FA would really be the minimum requirement. That particular amplifier, while relatively lightweight considering its specification, still comes in at a hefty 25kg which is more than many airlines will accept.

The SPE Expert 2K-FA amplifier is an HF + 6m amplifier with high output power but weighing in at 25kg.

Mention has been made of the desirability of running a beacon during a 6m DXpedition. This is not difficult to do. Many CW keyers have a beacon facility which is easily programmable. But most logging programs have a built-in CW capability which can also be programmed to beacon. Even the internal keyers of some transceivers can be set up to run a beacon.

As far as expedition antennas are concerned, again it comes down to what you are able to take. A key element of W6JKV's success on 6m expeditions has been the long-boom Yagi that he takes along. A good antenna not only gets out well but hears well, too, and may also be useful in nulling out local noise sources. So effort expended in getting a good antenna to your destination is generally effort well-spent.

The good news is that the hardware to get your antenna into the air can often be sourced locally once you arrive at your destination. Scaffold poles, ropes, pulleys and suchlike can be obtained from builders' merchants, ships' chandlers and similar stores which are found pretty much everywhere around the world. It may even be that local amateurs can help you out, too. The method of erecting a scaffold-pole mast using a gin-pole is shown in **Fig 8.1**. The mast typically consists of two or three scaffold poles (giving a height of 40ft or 60ft), with another pole for the gin-pole. The guys are 8mm polypropylene rope.

Co-axial cable may be harder to come by locally and this is something you may well need to check out before your trip. Hopefully you won't need too much. The sort of expedition that uses a scaffold mast attached to the roof rack of a hire car which doubles as the shack, for example, may only need a few metres of feeder, with the antenna sitting directly above the operating position. Bigger expeditions will plan differently – the UK-based Five Star DXers Association, for example, generally takes along a single long-boom Yagi or, occasionally, a stacked pair, which are put up on 40ft (13m) of scaffold poles and although the plans require the 6m antenna to be the nearest to the operating shack, the feeder run can easily be 100ft (30m) or so. But the team try to ship all their equipment in a 20ft shipping container, so that size and weight are not really an issue.

But do remember that coaxial cable becomes more lossy the higher the frequency. Multiband expeditions plan to have the 6m antenna system as the closest to the shack (shortest feeder run) and equip it with the best feeder they have available.

As for physically handling a large Yagi for 6m EME portable operation, various methods have been used to cater for the differing elevation angles required in order to track the moon. The T32C team, for example, were able to persuade the hotel's handyman to build a wooden support with hinge that allowed the Yagi to be tracked by pulling on and then tying down various ropes attached to the boom of the antenna. Somewhat 'Heath Robinson', but it worked!

Michael, DG1CMZ and Mike, G3WPH, congratulate each other after the first 6m EME contact made from T32C, Kiritimati (Christmas) Island, October 2011. The antenna is a 6M8GJ using manual azimuth and elevation control.

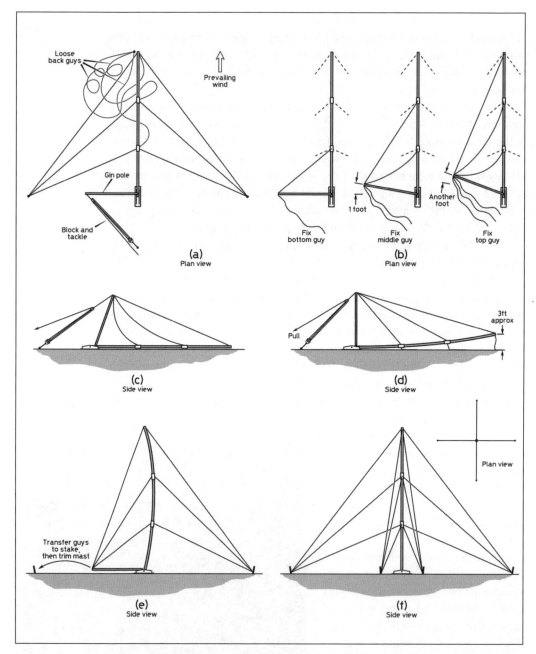

Fig 8.1: The gin-pole ('falling derrick') method of erecting a scaffold-pole mast. (a) Overhead view showing the mast and gin-pole laid out on the ground prior to erection. (b) Setting the pre-bend; the amount varies depending on the top load. (c) Side view showing the gin-pole being raised using the block and tackle. (d) As the guys take the strain, the mast should be slightly bent as shown. The antenna can now be fitted to the top of the mast. (e) With the mast in position, the gin-pole guys are transferred to stakes and adjusted to remove the bend (the other two sets of guys are not shown for clarity). (f) The mast in final position. Note the procedure should be reversed to lower the mast, ideally lowering away from the wind.

Whatever you plan, it doesn't usually make sense to use an antenna rotator. The weight of a rotator makes a tilt-over mast (which you will probably be using, unless you have access to some sort of telescopic push-up or pneumatic mast) very much more unwieldy. Usually it is much simpler to design the system so that it can be rotated by hand from the bottom. Remember to take along a compass to check the beam headings and then use stones, sticks or some other marker system at the base of the antenna to remind you of the main beam directions for North America, Europe, Japan etc.

GENERATORS / POWER SOURCES

One of the key considerations for a portable operation is power. If you are on some sort of Field Day operation with your local club this is unlikely to be an issue, as you will probably be able to take along a generator in the back of a car. But shipping generators or heavy duty batteries to the far side of the world is not to be recommended, unless they are part and parcel of a major DXpedition shipment. More likely you will want to source your power locally at your destination. Generators can be hired almost anywhere in the world and modern units are reliable, provide a well-regulated supply and are often quite quiet too. Battery power is fine for a 100-watt rig - a good car or truck battery will last for several hours, or more if you can charge it from your hire car. But battery power is unlikely to be sufficient if, as has been recommended, you use a linear amplifier. Ideally, you can find a rental villa that is well sited and has mains electricity, though even that is not necessarily the perfect solution that it may seem, as electricity supplies can be

Diesel or petrol generators can be more reliable – and are often less electrically noisy – than the mains supply in many countries. Here, a pair of Honda 3kVA petrol generators power four stations each running 500W amplifiers.

notoriously intermittent in some DX countries, with power outages inevitably occurring just as that much-awaited band opening comes along. Noise levels from mains power lines can also be a real problem in many countries.

Be sure to check what voltages will be available locally. Fortunately most modern equipment can be switched between 220 and 110V as required.

OTHER CONSIDERATIONS

Bear in mind other considerations such as health and safety issues, especially if your trip takes you a long way from the nearest medical facilities. They also include the need to take plenty of spares (connectors and suchlike) plus simple tools and test equipment (soldering facilities, multimeter, antenna test meter, for example), especially as you are likely to end up a long way from the nearest emporium. It isn't easy, especially if you have travelled by air, with the ensuing weight and other baggage restrictions and compromises will have to be made. But it must be so frustrating to be in the position that a failure occurs, such as the

It is not always necessary to utilise a massive antenna array to achieve success on 4m or 6m, as is well illustrated by this photo of the G7DDN/P station. Chris, G7DDN; Mark, G4FPH, and Dave, G3YXM, used a home-made 3-element Spiderbeam modelled in *EZNEC* and fashioned from bits of wire and broken fishing poles to win the RSGB 2010 6m Backpackers Contest.

5J0M problems mentioned earlier, and not to have the tools to hand to make at least an attempt at a repair.

Travel is actually one of the big limitations nowadays, in this age of worries about terrorism. Many officials are unfamiliar with amateur radio gear and may be suspicious of what you are carrying. Plenty of advice exists about how to handle this and it is worth thinking through beforehand such issues as what to carry in your hand baggage and what to check in (with the risk that it may arrive late at your destination or even go completely astray). If nothing else, do have your amateur radio licence to hand, along with any other relevant documentation, as proof of what it is that you are about.

DXPEDITION FUNDING

Finally in this chapter a quick word about expedition funding. DXpeditions will never be a way of making money, despite disparaging comments which occasionally appear on some Internet reflectors. They are expensive to mount and the only direct source of 'income', that is to say contributions sent with QSLs, is unlikely even to cover the cost of printing and posting the cards, especially if your expedition has been confined to one or two VHF bands, so that QSO totals are relatively modest.

If you consider the DXpedition as simply a form of activity holiday which you are happy to fund, then that is fine. But it is worth being aware that there are groups and foundations that are in a position to help with some of the radio-specific expenses of a legitimate expedition to a suitably rare location. You will be expected to cover your travel and accommodation costs, but bodies such as the UKSMG [5] or the Northern California DX Foundation (NCDXF) [6] are in a position, if a suitable case is made, to help with certain expenses, such as the cost of shipping an amplifier or antenna beforehand, or maybe offering the loan of a suitable lightweight transceiver.

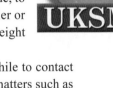

Whether you are asking for money or not, it is often worthwhile to contact such bodies before your trip, to find out other information about matters such as licensing, the rarity of your proposed location etc. They are often in a position to give useful advice.

REFERENCES

[1] *World Licensing and Operating Directory*, by Steve Telenius-Lowe, 9M6DXX, RSGB, 2008.

[2] 'Considerations for Successful 6m EME DXpeditions', Lance Collister, W7GJ; www.bigskyspaces.com/w7gj/DXPEDITIONS.htm

[3] Club Log: www.clublog.org

[4] G3WOS 6m amplifier: www.gare.co.uk/amplifier

[5] UK Six Metre Group: www.uksmg.org

[6] NCDXF, PO Box 1328, Los Altos, CA 94023-1328, USA; www.ncdxf.org

9 Contests, awards, QSLing

THIS CHAPTER COVERS a miscellany of topics that have been touched on elsewhere in the book and which will add to your enjoyment of Six and Four. Firstly, it covers some of the major contests on those bands. Whether or not you consider yourself a contester, contests generate activity and are therefore a great way to put contacts in your log including, all being well, some new squares, counties, or whatever it is you like to chase.

And that takes us nicely on to the second topic, awards. Because, whatever you do like to chase, there will almost certainly be an award that covers it, whether it be squares, countries, islands or states. There is lots of information on the Internet and elsewhere but this chapter covers some of the most popular operating awards as they relate to 6m and 4m.

If you want to chase awards, you will need to collect QSL cards and / or participate in the ARRL's Logbook of The World (LoTW) database, so the final part of the chapter covers these topics, albeit in fairly short order as there is plenty of material available from other sources.

CONTESTS

Contest operating has been mentioned already, in Chapter 5. Contests are an excellent opportunity to learn about propagation, to add to your totals, to build your operating skills and, of course, to have fun. One of the problems with 6m and 4m propagation, notwithstanding the excellent beacons that exist, is that the bands can be open somewhere from your QTH and you may be totally unaware of the fact because there is no activity at the far end. During a contest there are much greater levels of activity, giving a far better idea of band openings.

There are quite a number of contests which cover both bands, either as specifically 6m or 4m events or as part of a multiband contest. Most take place over the (northern hemisphere) summer period. 6m and 4m events are scheduled at that time because Sporadic E propagation is more likely to be about, making the contest more interesting. Multiband VHF events are scheduled for the summer because the higher VHF and UHF bands are essentially line-of-sight (most of the time) so portable operation from elevated locations is encouraged and it's not much fun setting up on mountaintops in the middle of winter!

The following paragraphs mention some of the scheduled 6m and 4m events in Europe and North America. These are subject to change and you should certainly check dates and rules from the relevant websites before packing your camping gear for the weekend or arranging for the XYL to be away so that you can do a serious contest effort from home.

In the UK, the main organisers of 6m contests are the RSGB and the UKSMG. The UKSMG Annual Summer Sporadic E Contest takes place in early June and is popular throughout Europe. The rules can be found on the UKSMG website and you don't need to be a UKSMG member to take part.

There are several RSGB contests that are of interest to both 6m and 4m operators. These include the following:

- 50MHz UK Activity Contest (every 4th Tuesday, January to November)
- 70MHz UK Activity Contest (every 5th Tuesday, January to November)
- 70MHz Cumulatives (February, March, May, June, August
- First 70MHz Contest (April)
- First 50MHz Contest (April)
- 70MHz CW (April)
- 50MHz Trophy Contest (June)
- 50MHz CW Contest (June)
- VHF National Field Day (July)
- 70MHz Trophy (July)
- Second 70MHz (September)
- 50MHz AFS Contest (October)
- 50/70/144/432MHz Christmas Cumulatives (December)

The dates, rules etc are published in the *RSGB Yearbook* [1] and on the RSGB VHF Contest website [2].

The *CQ* World Wide VHF Contest is held in July and includes 6m. This contest, although most popular in North America, also generates quite a lot of activity in Europe. Details from the *CQ* website [3].

The SMIRK QSO Party, sponsored by the Six Meter International Radio Klub, is held in June of each year. You do not need to be a SMIRK member to participate. Details from the SMIRK website [4].

The ARRL runs the following events which include 6m:

- ARRL January Sweepstakes (3rd or 4th weekend in January)
- ARRL June VHF QSO Party (2nd full weekend in June)
- Field Day (4th full weekend of June)
- ARRL September VHF QSO Party.

Details, not surprisingly, from the ARRL website [5].

The Four Metres Website lists almost thirty 70MHz events, several of which are local events (e.g. based in Scandinavia) although there is always the possibility that propagation may be favourable and contacts might be made with other areas.

There are no doubt quite a few other 6m and 4m contests around the world and if your national society doesn't sponsor anything appropriate, you might suggest that it does so. Here are a few others: the OZ 50MHz Open Cumulative Contest runs on the fourth Tuesday every month [6], and the SARL (South African Radio League) run several contests which include 50MHz [7]. The UK magazine *Practical Wireless* also sponsors an annual 70MHz contest which remains popular, especially given the upturn in 4m activity in recent years.

In most respects contesting on Six and Four is no different to contesting on other bands. If you are taking part in the contest simply to find DX such as new squares and countries by way of all that contest activity, then all you need to do is tune the band and work them or, if you prefer, use the Internet-based facilities that have been discussed elsewhere such as the *DX Summit* website and the ON4KST chat room to be alerted to any particularly interesting contest activity. You may also want to run a local CW Skimmer or log into the Reverse Beacon Network. Be aware that, if you intend putting in an entry, the use of external spotting facilities may well put you in an Assisted or Multi-Operator category (some contests rule out such assistance completely, which

The single-operator 6m contest station of Roger, G4BVY/P, in the RSGB 50MHz contest in April 2013. The station took about 30 minutes to set up and consisted of an Elecraft K3 transceiver running from the Land Rover battery to an F9FT 5-element Yagi, showing that a simple station can be easily set up while also being reasonably competitive. The location was a hilltop about 10km from Malvern.

would then require you to submit as a checklog).

Before working any contest stations, at the very least check the contest exchange (often simply RS(T) plus locator square, but sometimes including a serial number, UKSMG membership number or other data), so that when you work someone you know immediately what he needs from you. You can probably determine the exchange from listening to a few QSOs, but the best way is to check out the rules via the organiser's website. Any rare stations in the contest (particularly if they are out portable to activate an unusual grid square) will want to amass as many contest QSOs as possible, so contests are not the time for exchanging names or other extraneous information.

If you plan to take part seriously in contests, then rather more effort is required, both before and during the contest. Beforehand, check the rules carefully and plan your strategy based on them, especially the scoring system. Does it favour distance, multipliers (e.g. grid squares) or some other aspect? Check out your station, too, to ensure that there are unlikely to be any technical problems while the contest is underway. There's quite a bit more you can do, too, such as preloading the standard messages into your voice or CW keyer. Warn the family that you will be unavailable, think about your food and sleep strategies, and so on. All of this is familiar to contesters, whatever the band. If you plan to operate portable, maybe to take advantage of a hilltop location, then there is plenty of other preparation to do, covered in Chapter 8. Don't forget, for example, to check the grid square of your portable location and print it up on a card so that you don't accidentally send your home station grid during the contest! Check also, via the various bulletins or websites, what portable and other expedition activity will be taking place during the contest, so that you know what to be on the lookout for.

So how about aspects that are specific to Six and Four? Having read this far, you will be fully aware that these bands are noted for the many propagation modes that they enjoy. So you will want to think about what propagation is likely during the contest (many VHF contests are, of course, scheduled for the Sporadic E season) and be ready to make the most of any band openings which occur. If, for example, the contest scoring is distance-based, you simply cannot afford to miss any fleeting openings even if it means passing up on a nice run of more local contacts. And if, say, there is a multiplier system based on squares and / or DXCC entities, then you will want to ensure that these go into the bag even if, again, you have to take time out from a run of more mundane stations to do so. It's your choice whether you use Internet-based assistance or not. As mentioned, some contests allow it, some allow it but put you in a different category, some won't allow it at all, in which case you can still use such assistance but if you send in your log it will be treated as a checklog. But there is no restriction on local alerting techniques such as a separate receiver monitoring any commercial frequencies that might alert you to possible band openings or the use of a radio with a spectrum display so that, while working stations on your frequency, you can get an idea of activity elsewhere in the band.

If the contest runs for, say, 24 hours, you will want to schedule any rest periods at times when other stations are least likely to be active and when propagation is likely to be poorest. This will almost certainly mean the middle of the night, when you are sure that any Sporadic E has finally died away.

There is no magic secret to becoming a successful contest operator, except to take part in as many events as possible so that the actual operating becomes second-nature and your brain can focus on the strategic aspects such as when to check different directions, when to sweep the band and when to CQ or when to switch modes (in multimode events). And during the quiet times you may even find it profitable to drum up some contacts from local stations operating FM simplex (repeater contacts are unlikely to be valid for contest points). Every contest is different and this is what makes them fun. You may be plodding away at a low QSO rate and getting frustrated, ready to take a coffee break just at the moment that the band explodes with Sporadic E propagation and your log starts to fill up with distant callsigns. When rates are slow you may well be happy to chew the fat a little with other stations you work, just to stay awake and alert. When it gets busy, the good operator will be ready to get his contest exchange away in a snappy fashion and copy the other station's exchange first time, without needing to ask for repeats. Whether you record the whole contest to your hard disc to be able to listen again later is up to you. Some contesters consider this against the spirit of contesting while others would not be so concerned. The general consensus is that it is fine to make a recording for later analysis and interest, but not appropriate to use it for amending the contest log before submission.

To summarise, the main advice is to check out the rules beforehand, and think carefully about which category you intend to enter, your strategy to achieve a successful result etc. What you consider a successful result may vary. You may be interested in using the contest to add to your total of squares worked, in which case you may choose to use 'assistance' during the contest, by way of *Cluster* or the ON4KST chat room. But some contest organisers will then only accept your log as a checklog. This is fine if you have met your goals, but if you want to make a serious entry check the rules first and if, for example, those facilities are ruled out for competitive entrants bear that in mind in your planning. If you are heading out portable for a contest the usual pre-planning is necessary.

And do send in your log afterwards, whether as a competitive entry or as a checklog (in the format specified in the rules, please). It will be appreciated by the contest organisers as a way of checking other entrants' logs and the contest database as a whole may even be helpful for later analysis as regards propagation mechanisms at work over the contest weekend.

AWARD CHASING

Award chasing is a popular pastime on all bands. Many of the major awards allow single-band endorsements or have a specific 6m or 4m version of the award. First and foremost is the **ARRL DXCC (DX Century Club) award**. The 6m version of DXCC was introduced in 1990 and eight amateurs qualified that year (contacts from 1945 onwards were eligible). At the time of writing, over 1000 amateurs had

The author's 6m DXCC certificate, with endorsements for 125 and 150 entities.

This DXCC Challenge plaque is available to those who have confirmed at least 1000 current DXCC credits on HF plus 6 metres.

received this award, which is for working and confirming a minimum of 100 DXCC entities from the overall list of 340. This is a tough achievement if you live somewhere like the US West Coast, whereas European amateurs have a head start, with something like 60 DXCC entities on their doorstep or, at least, within annual Sporadic E range.

Full details of how to claim DXCC are available from the ARRL website. The good news is that, since the advent of Logbook of the World you may not need to collect traditional QSL cards for the award application. If the stations you work submit their logs to LoTW you can claim credits electronically. You may still want to collect QSL cards for the shack wall, but at least you don't have to risk sending them through the mail for checking.

A lists of early recipients of the 6m DXCC can be found on ZS6EZ's website [8] and the current list of all holders, with their DXCC standings, can be downloaded as a PDF file from the ARRL website [9].

An extension to DXCC is the ARRL 'DXCC Challenge' award, which is earned by working and confirming at least 1000 DXCC band-points, using any bands from 160 to 6m (except 60m). Those with more than 1000 band-points can claim a handsome wooden plaque, which is endorseable in increments of 500. The inclusion of 6m in the DXCC Challenge has encouraged many HF DXers to take a keen interest in Six too!

WORKED ALL STATES

As well as DXCC, LoTW also supports the Worked All States award. WAS is certainly an achievable goal for North American and Caribbean stations active on 6m, but perhaps somewhat more challenging from, say, Europe or Japan. Single-band WAS awards are available for 6m operation. Application and record forms are available from the ARRL website.

VUCC

Another ARRL award, specifically geared to VHF operation, is the VHF / UHF Century

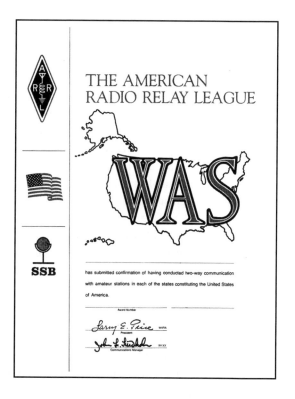

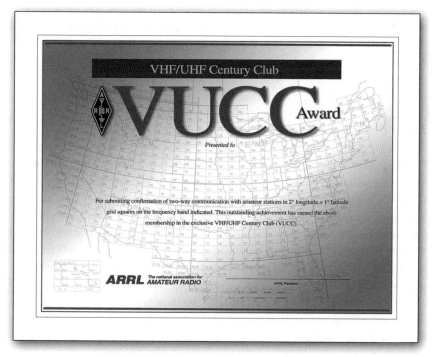

The VUCC Award, issued by the ARRL.

Club award (VUCC). VHF / UHF operators have tended to want to chase some sort of geographical entity smaller than a DXCC country, because high country totals are tough on VHF, especially on 2m and above. The popular answer is the chasing of Locator Squares which were introduced originally as a way of pinning down station location for the scoring of VHF contests.

The original five-character QRA locator system, based on latitude and longitude, has its origins in Europe back in 1959, but was unsuitable for world-wide use as grid references would not be unique but would repeat at intervals. Therefore, in 1980, a VHF working group meeting in Maidenhead, UK, adopted a proposal by John Morris, G4ANB, based on a newly generated world-wide system devised by Folke Rasvall, SM5AGM.

This system covers the world in grid locators which are 2 degrees of longitude in width and 1 degree of latitude in height. If this seems odd it is useful to think that, at the equator, they will be twice as wide as they are high, but at the poles their width will be zero! At the latitudes where the bulk of the amateur population resides, they are not far off square (this system actually makes grid square chasing in, say South Africa, much tougher than in, say, northern Europe, because each locator 'square' actually has a much greater area in ZS). These basic grid squares are designated by two letters and two numbers (e.g. IO91), though a further two letters can be added to give a more precise location. More information on the grid square system can be found in **Appendix F**.

Locator square chasing is very popular on all VHF / UHF bands, not least on 6m. The rules of the VUCC award appear on the ARRL website. The basic requirement for a 6m award is 100 credits, which means 100 confirmed squares worked since 1 January 1983. In reality, from pretty much anywhere in Europe or North America, 100 grids should be workable in a matter of months, or even weeks, even with a very modest station, especially if you focus on those periods of the year when Sporadic E is common.

Fred Fish Memorial Award (FFMA)

In 2008 the ARRL Board of Directors approved a new award honouring the late Fred Fish, W5FF, the only amateur who worked and confirmed all 488 grid squares in the 48 contiguous United States on 6 metres (and also the second recipient of 6m DXCC). The ARRL press release announcing the award states, "The Fred Fish Memorial Award will be granted to any amateur who duplicates Fish's accomplishment. Fish was a mainstay on the VHF+ bands for many years, having achieved Worked All States (WAS) on 6 metres through 432MHz, as well as DXCC for 6 metres. He is widely regarded as a gentleman operator and one of the finest amateurs in the VHF+ community. ARRL Contest Manager Sean Kutzko, KX9X, will oversee this award. Kutzko, an avid VHF+ operator himself, said, 'We hope the new award will increase 6 metre activity throughout the US and the world. We also hope it will lead to the activation of rare grid squares in the US by encouraging the native ham population of a rare grid square to give 6 metres a try, as well as through so-called Grid DXpeditions. We actively call on the 6 metre community to help educate VHF+ newcomers to the fun that is available on 6 metres'."

Complete details on the Fred Fish Memorial Award are available from the ARRL website.

Incidentally, for all ARRL awards amateurs in the USA and US possessions need to be ARRL members. Amateurs elsewhere in the world may apply without being ARRL members.

50MHz WAZ

CQ Magazine issues the Worked All Zones awards for contacts with the 40 zones into which the world is divided for purposes of this award. A single-band award is available for 6m operation, Mixed Mode only, and is issued for a minimum of 25 zones confirmed, with endorsement stickers for 30, 35, 36, 37, 38, 39 and 40 zones. Contacts must have been made since 1 January 1973. A list of 'top contenders' showing zones still needed appears on the *CQ* awards website [10]. At the time of writing LZ1CC needed just Zone 1 to complete all 40 zones, while JF1IRW and JA7QVI were each short of zones 2 and 40. In all, 113 6m DXers had qualified for the award.

RSGB AWARDS

The RSGB Islands on the Air (IOTA) awards programme has a VHF / UHF version, which is increasingly popular with 6m operators. Full details are available from the RSGB IOTA website [11].

The RSGB also issues 50MHz and 70MHz Countries and Squares awards, the

The RSGB Islands on The Air (IOTA) scheme, very popular on the HF bands, is also available for 6m enthusiasts and is attracting a growing following.

RSGB 50MHZ AWARDS

There are three awards for the 50MHz band and all are dependent to some extent on the special propagation modes that can be experienced at these frequencies. Both squares and countries are catered for, but independently, and as a reminder of the potential for cross-band working, there is a special DX certificate.

50MHz Countries Award

The initial qualification for this certificate is proof of completed two-way QSOs on 50MHz with 10 countries. Only contacts with countries permitting 50MHz operation can be considered. Stickers will be provided for increments of every ten countries worked.
RULES:
All contacts must have been on or after 1 June 1987.
QSL cards submitted must be arranged in alphabetical order of the countries claimed, and a checklist enclosed.
Stations are eligible for awards in the following categories: (i) Fixed stations; (ii) Temporary location or portable (/P) operation (categories cannot be mixed).

50MHz DX Certificate

This certificate takes into account the considerable potential for cross-band working when transmitting in the 50MHz band. There is therefore no stipulation on the band used for the incoming signal.

The initial qualification is confirmation from 25 different countries of a successful QSO with transmission from the applicant's country taking place within the 50MHz band. Stickers will be provided for increments of 25 countries confirmed.
RULES:
All contacts must have been on or after 1 June 1987.
QSL cards submitted must be arranged in alphabetical order of the countries claimed, and a checklist enclosed.
Stations are eligible for awards in the following categories: (i) Fixed stations; (ii) Temporary location or portable (/P) operation (categories cannot be mixed).

50MHz Squares Award

The 50MHz Squares Award is intended to mark successful VHF achievement. The initial qualification needed for this certificate is proof that 25 different locator squares have been worked with complete two-way QSOs within the 50MHz band. Squares in any country will qualify, provided that operation from that country is formally authorised. Additional stickers will be provided when proof is submitted for increments of 25 squares.
RULES:
All contacts must have been on or after 1 June 1987.
QSL cards submitted must be arranged in alphabetical order of the QTH squares claimed, and a checklist enclosed.
Stations are eligible for awards in the following categories: (i) Fixed stations; (ii) Temporary location or portable (/P) operation (categories cannot be mixed).

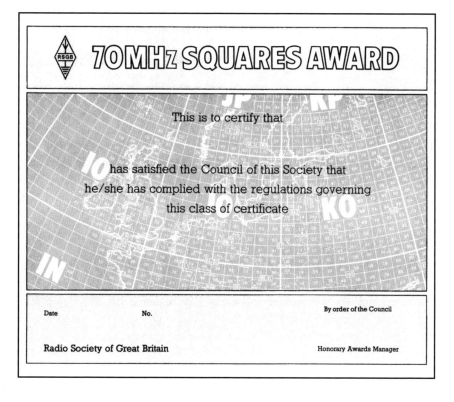

Above: Samples of two RSGB awards for 6m and 4m. Brief details of these and other RSGB 6m / 4m awards are given in the sidebars.

RSGB 4m AWARDS

Squares and Countries

Part of the traditional '4-2-70 Squares Awards', these mark successful achievement on these bands. Initially, a certificate and one sticker will be issued. Further stickers will be issued as additional locator squares are claimed. The title of each award gives the number of locator squares and countries needed to qualify for the award. For example, to obtain the basic 70MHz 20/4 award you must have QSL cards confirming contact with 20 locator squares including 4 countries on 70MHz. Endorsement stickers are available for up to 55 squares and 10 countries.

4m Countries and Postal Districts

Awarded for confirmed contacts with:

Standard Award	45 districts	3 countries
Senior Award	80 districts	6 countries

A list of the eligible postal districts and countries can be found in the *RSGB Yearbook.*

latter as part of the '4-2-70' awards programme. Further details are shown opposite and are also available in the *RSGB Yearbook* as well as on the VHF awards page of the RSGB website [12].

UKSMG AWARDS

The UKSMG issues a number of awards for 6m operation. There are continental awards, for example for working 10 DXCC countries in the continent of Asia, a Digital award for working grid squares using one of the digital modes (RTTY, JT6M etc) and various awards for increasing levels of achievement on 6m EME. Full details available from the UKSMG website.

WSJTGROUP AWARDS

The *Yahoo*-based WSJTGROUP sponsors several awards for QSOs made specifically with WSJT. The most popular is the *10,000 Miles Award*, which is presented to any amateur who completes the basic requirements and submits their log to the awards committee. The objective is to complete two-way meteor scatter contacts with other stations; the distance between stations is the basis for the accumulated miles. The minimum distance between stations must be 1000 miles and the total accumulated must be 10,000 miles or more. Endorsements are issued in multiples of 5000 miles once a station has completed the basic award. There is no charge for the award: a PDF certificate is e-mailed to recipents.

The *WSJT HSMS Initial QSO Award* is to acknowledge the accomplishments of Meteor Scatter operators chasing 'Initial' contacts. An initial contact is either a different callsign or the same callsign that travels to a different four-digit grid square or state or DXCC entity, even if that State or DXCC entity is within the original four-digit grid square. Note that the rules for initials are not the same as for EME initials. The starting point is 25 Initial Contacts. Endorsements are available for higher levels.

The *Random QSO Award (RQA)* is earned by completing 15 random contacts during Random Hour. Endorsements are available for the RQA. The RQA promotes the proper use of the calling frequency for group 'Activity Periods'.

Details including full rules and information on how to apply appear on the WSJTGROUP official website [13].

* * * * *

The above list of awards is by no means intended to be comprehensive and, in any case, many general operating awards can be obtained with a single-band endorsement for 6m. For a comprehensive source of information about amateur radio awards, the best single resource is the K1BV website [14].

Neither an award nor a contest, but a brief mention of the Six Metres Marathon [15] might be in order here. This is run annually by Hannu, OH3WW / OH1HS, and runs from May to August, in other words the main 6m DX period of the year. A great incentive to chase 6m DX while you are waiting for that elusive 'new one' to appear.

QSLING

Some amateur radio awards work on an 'honour' basis; in other words, your application simply needs a signed declaration that you have worked the stations concerned (or that you have confirmations of those contacts). Some require a third party to see your cards and endorse your application (typically the rules may require that your application is endorsed by two officials of a local radio club). Others require that you send QSL cards to the organisation that actually issues the award.

Whatever the circumstances, and even if you don't chase awards, it is always nice to have a QSL card as a permanent record of a memorable contact. The arrival of the QSL is also reassuring if the contact was made under marginal circumstances and, perhaps, you weren't 100% sure that the contact was OK!

The QSL bureau system has worked effectively for many years, but it can take a matter of years for your card to reach

Peter Lund, LA7QIA, first operated on 6m as JW7QIA in 1996. In 2010 he activated 4m for the first time from Svalbard. A superb QSL on 6m or 4m in anyone's collection!

the distant end and a card to find its way back to you. That's fine for the bulk of contacts that you make in day to day operation or in a contest, but for that rare one that you need for an award or to put on the shack wall you will want a somewhat quicker turnaround. In any case, within the USA the ARRL bureau only handles cards for overseas contacts, so any domestic cards you need for, say, WAS, you will have to chase directly via the postal service. Most DX operations and DXpeditions appoint a QSL manager to speed the process up and, in some cases, to get over the problem of postal theft which is common in many countries.

Lots of guidance has been offered on how to do direct QSLing to maximise your chances of a return card and I won't repeat it here. Check sources such as the *RSGB Operating Manual* for detailed advice. But generally it is a matter of common sense, such as including an addressed return envelope with funds for return postage (usually a couple of dollar bills or an International Reply Coupon) along with your QSL card on which you have correctly entered all the QSO information (remember, for example, to put times in UTC, not local). And if sending overseas, it's generally wise not to put any reference to amateur radio on the outside of the envelope or any suggestion that it may contain something of value.

If you are planning on having QSL cards printed, ensure that your card carries information that is of importance to other 6m and 4m operators, i.e. your country, state (in the USA) or county (in the UK) and, most particularly, your grid square. You may also want to add equipment and antenna details: many 6m enthusiasts have two designs of card printed; one specifically for their 6m QSOs and one for everything else!

T30SIX was the callsign of the dedicated 6m station on a Brazilian DXpedition to Western Kiribati in October 2012.

LOGBOOK OF THE WORLD

The ARRL's Logbook of the World (LoTW) is now well-established and supports the DXCC and WAS awards programmes. You can apply for DXCC awards and endorsements using a mixture of traditional QSL cards and matches on LoTW. The latter requires that you upload your own log to LoTW and the system will match any contacts it is able to with other logs on the LoTW system. Increasingly, DXpeditions are uploading their logs to LoTW as a matter of course, so this can be a convenient and inexpensive way of getting those credits for DXCC. The ARRL is understood to be enhancing the LoTW software to enable links to other popular awards programmes (for example, *CQ Magazine's* WPX awards are now supported)), but this is by no means trivial while protecting the integrity of the LoTW database, so don't expect immediate developments on this front (at the time of writing, ARRL was conducting a consultation exercise on the future direction of LoTW).

eQSL

While on the topic of QSO databases for awards purposes, it is worth mentioning that there are other such databases in existence, the most popular of which is eQSL [16]. eQSL does not have the wraparound of public key encryption that is fundamental to LoTW and is therefore potentially more open to abuse. But if you simply to want to see matches to your contacts and print attractive cards for the shack without the cost of direct QSLing, you may want to give eQSL a try. It is certainly true that many amateurs nowadays (particularly of the younger generation) see no role for 'traditional' paper QSL cards and would prefer to have all their confirmations in electronic form with, perhaps, the option to print out a confirmation locally (which eQSL offers).

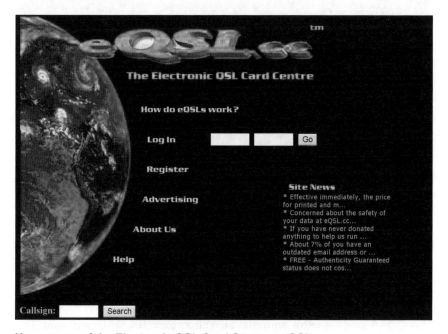

Home page of the Electronic QSL Card Centre – eQSL

OQRS (ONLINE QSL REQUEST SERVICE)

Perhaps the major change in QSLing since the *6 Metre Handbook* was published in 2008 has been the widespread acceptance of Online QSL Request Services (OQRS). This is a logical development for a number of reasons. The basic premise is that you can order a direct or bureau card via an Internet site rather than have to mail a physical card with the associated cost, risk of it being lost *en route* and inevitable delays in the post and / or the bureau system. It also has the advantage that it is never the best idea to send cash (such as dollar bills) through the mail, while IRCs (International Reply Coupons) are being phased out by many postal administrations (signatories to the international postal conventions are still obliged to honour IRCs, but many no longer make them available for sale).

OQRS started with a number of DXpeditions making a request facility available on their website, usually in association with a capability to make a donation via *PayPal* (or other means) to cover mailing costs and perhaps help to cover other DXpedition costs. This move made a lot of sense – very few DXpeditions wish to retain incoming QSL cards so why collect them in the first place?

Many expeditions continue to offer OQRS via their own websites, but a powerful and user-friendly alternative has come to the fore in recent years by way of G7VJR's excellent *Club Log* [17]. *Club Log* offers not only expeditions but also individual DX stations and, indeed, any user the ability to receive QSL requests via the site. So, for example, if you don't wish to collect QSL cards for your more

The **OQRS page on the website of well-known UK QSL manager Tim Beaumont, M0URX.**

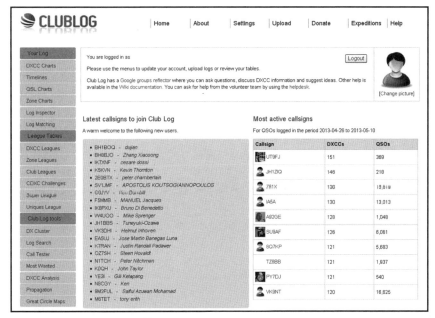

The home page of *Club Log* by G7VJR.

mundane day-to-day contacts, you can still receive and honour QSL requests from those who would like to have a card from you.

Club Log offers many other facilities too, as has been mentioned elsewhere in this book and new features are being added on a regular basis. It is well worth signing up and taking advantage of the facilities it offers.

REFERENCES

[1] *RSGB Yearbook*, published annually by RSGB.

[2] RSGB VHF Contests: www.rsgbcc.org/vhf

[3] *CQ* WW VHF Contest: www.cqww-vhf.com

[4] SMIRK: www.smirk.org

[5] ARRL Contest Calendar: www.arrl.org/contest-calendar

[6] OZ VHF Contests: www.qsl.net/oz6om/the50mhzopen/The50MHzOpen.html

[7] ZS VHF Contests: www.sarl.org.za/public/contests/contestrules.asp

[8] ZS6EZ list of 6m DXCC recipients: http://zs6ez.org.za/lists/dxcc-6m.txt

[9] ARRL list of DXCC holders and scores: www.arrl.org/dxcc-standings

[10] *CQ* awards: www.cq-amateur-radio.com/cq_awards/index_cq_awards.html

[11] IOTA: www.rsgbiota.org

[12] RSGB VHF Awards: www.rsgb.org.uk/vhfawards

[13] WSJTGROUP official website: www.meteorscatter.org

[14] K1BV awards site: www.dxawards.com

[15] 6m Marathon: http://tamrinki.fi/6m/mindex.php

[16] eQSL: www.eqsl.cc/qslcard/Index.cfm

[17] Club Log: www.clublog.org

Appendices

APPENDIX A – GLOSSARY
OF TERMS AND ABBREVIATIONS

AM	Amplitude Modulation
ARRL	American Radio Relay League (the national body represent ing amateur radio in the USA)
CEPT	The European Conference of Post and Telecommunications Administrations
CW	Continuous Wave (refers to Morse transmissions)
DSP	Digital Signal Processing
DX	Long-distance (or rare) in the context of stations contacted
DXCC	DX Century Club (major international awards programme)
DXing	Chasing contacts with rare or distant stations (see DX)
DXpedition	An expedition specifically to activate a DX location
EMC	Electromagnetic Compatibility
EME	Earth-Moon-Earth, using the moon as a passive reflector to make long-distance contacts
ERP	Effective Radiated Power
FCC	Federal Communications Commission (US licensing body)
HF	High Frequency (officially 3 - 30MHz, but used by amateurs to refer to the bands 30 to 10m)
IARU	International Amateur Radio Union
IOTA	Islands on the Air (major RSGB Awards Programme)
ITU	International Telecommunication Union
LoTW	Logbook of The World
lpm	Letters per Minute
MS	Meteor Scatter
MUF	Maximum Usable Frequency
NCDXF	Northern California DX Foundation.
NoV	Notice of Variation (to UK amateur licence)
OFCOM	Office of Communications (UK)
PC	Personal Computer

QSL Card	Card exchanged by radio amateurs to confirm a successful two-way contact
QSO	Used to refer to a contact between radio amateurs
RSGB	Radio Society of Great Britain
sked	Pre-arranged schedule
SMIRK	Six Metre International Radio Klub
SSB	Single Sideband
SWR	Standing Wave Ratio
TEP	Trans-Equatorial Propagation
UKSMG	UK Six Metre Group
UTC	Universal Coordinated Time (effectively the same as Greenwich Mean Time)
VHF	Very High Frequency (officially 30 - 300MHz, used by amateurs to refer to the 50, 70, 144 and, in North America, 220MHz bands)
WAS	Worked All States (an awards programme)
WAZ	Worked All Zones (an awards programme)
WSJT	Weak Signal by K1JT (a suite of programs developed for weak signal working)
Yagi	The most popular type of directional antenna (named after one of its Japanese inventors)

APPENDIX B – SOURCES OF INFORMATION

This Appendix lists a wide range of sources, including those referenced elsewhere in this book, where you can find information, software to download, and much else that is relevant to the subject matter of the book. Inevitably, the longest list is of websites, as these offer the most up-to-date sources and are easily accessible (and free!) However, it must be remembered that websites come and go, and their addresses change, but the list is believed to be accurate at press time. A set of links, including those in this Appendix, will also be maintained via the author's website (www.g3xtt.com, and follow 6m book and subsequent links). Readers are welcome to send in their own suggestions (to don.field@gmail.com) for inclusion on that site and for possible inclusion in the next edition of this book.

GENERAL (HISTORY, CLUBS ETC)
Books
VHF / UHF Handbook, RSGB.

RSGB Yearbook, published annually by RSGB.

The Amateur Radio Operating Manual (7th edition), Edited by Don Field, G3XTT, and Steve Telenius-Lowe, 9M6DXX, RSGB 2010.

The ARRL Handbook, ARRL.

Radio Communication Handbook, RSGB.

The VHF / UHF DX Book, G3SEK and others, DIR Publishing, 1992 (republished by RSGB but now out of print).

The ARRL Operating Manual for Radio Amateurs, 10th Edition, ARRL.

ARRL's VHF Digital Handbook, Steve Ford, WB8IMY, ARRL.

Articles
'What to Expect on 6', Bill Wageman, K5MAT, *QST* August 2004, pp 49-51.

Websites
SMIRK: www.smirk.org

JI1CQA: http://ji1cqa.sakura.ne.jp/index.shtml

JA1RJU ('History of VHF in Japan', in English): www5.big.or.jp/~ja1rju/history_japan.html

EY8MM: www.ey8mm.com

PY5CC: www.50mhz.com/py.html

VK3OT tape archives: www.telesupport.nl/wave/vk3ot/index.htm

M5BXB: www.m5bxb.com

Six Club (Six Meters World Wide): http://www.6mt.com

DK7ZB: www.mydarc.de/dk7zb/start1.htm

OE 6m Pages (lots of 6m links): www.qsl.net/oe4whg/6mlinks.htm

DX Zone: www.dxzone.com/catalog/Operating_Modes/50_Mhz

DK5YA VHF Page: www.vhfdx.de/6mtrs.html

UKSMG: www.uksmg.org

VK3SIX: http://home.vicnet.net.au/~vk3six

G3WOS: www.gare.co.uk/g3wos.htm

VK4CP: www.qsl.net/vk4cp

H44PT (G0BCG): www.h44pt.org.uk

PROPAGATION

Books

Radio Auroras, Charlie Newton, G2FKZ, revised edition with new chapter by Neil Carr, G0JHC, RSGB 2012.

Sun, Earth & Radio, J A Ratcliffe, World University Library, 1970 (out of print).

Ionospheric Radio, Davies, ISBN 0 86341 186 X.

Articles

'Sporadic E – A Mystery Solved?' Parts 1 & 2, *QST*, October & November 1997.

'Making the Most of Sporadic E at VHF', *RadCom*, January 2002, p34.

'Es Sporadic E propagation at VHF: a review of Progress and Prospects', Emil Pocock, W3EP, *QST* April 1988.

'Meteor Scatter Communications', Clarke Greene, K1JX, *QST*, January 1986, pp14-17.

'Catch a Falling Star', Kirk Kleinschmidt, NT0Z, *QST*, October 1997, pp63-67.

Websites

50MHz F2 Propagation Mechanisms: www.ham-radio.com/n6ca/50MHz/K6MIO_50MHz_F2Prop.pdf

A Seven Year Study of 50MHz Propagation: www.qsl.net/w/wa5iyx/cq/cq7208a.htm

Useful references and free software: www.sss-mag.com/prop.html#software

'Mid Latitude Sporadic E, A Review' (Michael Hawk), www.amfmdx.net/propagation/Es.html

Solar weather – NOAA Solar Data: www.noaa.gov

Solar Ham by VE3EN: www.solarham.net

DF5AI: www.df5ai.net

Space Weather: www.spaceweather.com

K1SIX: http://k1six.com/6M_Es.html (with links to other good sites)

RSGB Propagation Studies Committee: www.rsgb.org.uk/psc

International Meteor Organisation: www.imo.net

G3USF 50MHz Beacon List: www.keele.ac.uk/depts/por/50.htm

Propagation: www.sss-mag.com/prop.html#software

Meteor Scatter: www.dxzone.com/catalog/Operating_Modes/Meteors

QST TEP article, part 1: www.vhfdx.net/docs/qst_te_nov_1981_part1.pdf

QST TEP article, part 2: www.vhfdx.net/docs/qst_te_dec_1981_part2.pdf

TEP bibliography: www.dxmaps.com/tepbiblio.html

TEP article: www.ips.gov.au/Category/Educational/Other%20Topics/Radio%20Communication/Transequatorial.pdf

OPERATING
Websites

ZS6EZ website: http://zs6ez.org.za

G3WOS list of UK 'firsts': www.gare.co.uk/ukalltime.htm

List of 6m allocations: www.secornwall.pwp.blueyonder.co.uk/licensing.pdf

ON4KST: www.on4kst.com/chat/start.php

The Daily DX: www.dailydx.com

OPDX Bulletin: www.papays.com/opdx.html

425 DX News: www.425dxn.org

DX Summit: www.dxsummit.fi

DXLite: http://dxlite.g7vjr.org

WinSDR: http://psn.quake.net/WinSDR

CW Skimmer: www.dxatlas.com/CwSkimmer

QTH Locator: http://f6fvy.free.fr/qthLocator

DX World: http://dx-world.net

EQUIPMENT & ANTENNAS
Books

The ARRL Antenna Book, ARRL.

Antenna Compendium No.5, ARRL ('Two portable 6-Meter Antennas', designs for 2-ele quad and 3-ele Yagi).

ARRL's VHF / UHF Antenna Classics, ARRL

Twenty Five Years of Hart Reviews, Peter Hart, G3SJX, RSGB.

RSGB Antenna File, compiled and edited by Steve Telenius-Lowe, 9M6DXX, RSGB, 2013.

Article

'Six Metre Transceivers', *SixNews*, Issue 65, May 2000.

Websites

Amplitec: www.amplitec.hu/ug_4_100_1000_gs31b_eng.html

FT-847 mods: http://oz1djj.geronne.dk/ft847_4m_mods.htm

6m & 4m antenna designs: www.yu7ef.com

Remote Control via Internet: www.dh7fb.de

G3WOS 6m amplifier design: www.gare.co.uk/amplifier/index.htm

RF Hamdesign: www.rfhamdesign.com/products/powersplitters/index.html

GM3SEK: www.ifwtech.co.uk/g3sek

Cable loss data: www.qsl.net/dk3xt/cable.htm

Transmission line loss calculator: http://vk1od.net/calc/tl/tllc.php

U310 Preamp: www.ham-radio.com/n6ca/50MHz/50appnotes/U310.html

Low Pass Filter: www.ham-radio.com/n6ca/50MHz/50appnotes/50tlpf.html

Remote control: www.n0hr.com/hamradio/45/10/ham_radio0.htm

WEAK SIGNAL WORKING
Articles
'I Don't use Digital, Honestly I Don't', Ken Osborne, G4IGO, *SixNews*, Issue 94 (Feb 2008).
'EME with JT65', Gene Zimmerman, W3ZZ, *QST*, June 2005, pp80-82.

Websites
G0CHE website: www.g0che.co.uk/home.php
WSJT Page: http://physics.princeton.edu/pulsar/K1JT
WA5UFH: www.qsl.net/wa5ufh
JT6M: www.jt6m.org/home.php
Dimension4: www.thinkman.com/dimension4/download.htm
Ping Jockey: www.pingjockey.net/cgi-bin/pingtalk
International Meteor Organisation: www.imo.net
Virgo Meteor Sky View (DL1DBC): www.dl1dbc.net/Meteorscatter
OH5IY: www.kolumbus.fi/oh5iy
W7GJ: www.bigskyspaces.com/w7gj
JT65 Terrestrial chat room: www.chris.org/cgi-bin/jt65talk
GM4JJJ: www.gm4jjj.co.uk/MoonSked/moonsked.htm
IW5DHN: www.qsl.net/iw5dhn/when.htm
N1BUG: www.g1ogy.com/www.n1bug.net/prop/eme.html
DF9CY: www.df9cy.de/ar/radio.htm
VK3UM EME software: www.vk3um.com
SM2CEW EME: www.sm2cew.com

AWARDS / QSLING
Websites
V-U-SHF site: www.vushf.dk
ARRL list of DXCC holders and scores: www.arrl.org/awards/dxcc
IOTA: www.rsgbiota.org
RSGB VHF Awards: www.rsgb.org.uk/vhfawards
K1BV awards site: www.dxawards.com
eQSL: www.eqsl.cc/qslcard/Index.cfm

CONTESTS
Websites
RSGB VHF Contests: www.rsgbcc.org
CQ WW VHF Contest: www.cqww-vhf.com
ARRL Contest calendar: www.arrl.org/contests/calendar.html
OZ VHF Contests: www.qsl.net/oz6om/the50mhzopen/The50MHzOpen.html
ZS VHF Contests: www.sarl.org.za/public/contests/contestrules.asp

FOUR METRES

Website

4m website: www.70mhz.org

4m linear amplifier: www.amplitec.hu/ug_4_100_1000_gs31b_eng.html

OZ2M 70MHz transverter design: www.rudius.net/oz2m/70mhz/transverter.htm

UK 4m reflector: fourmetres@yahoogroups.com

APPENDIX C – UK 6M REPEATERS

Callsign	6m channel	Output freq	Input freq	Loc	Location	CTCSS	Keeper	Status (see Notes)
GB3WX	–	50.52	29.21	IO81VC	S Wilts	77.0Hz	G3ZXX	1, 5
GB3AE	R50-01	50.72	51.22	IO71PR	Tenby	94.8Hz	GW4AKZ	1
GB3EF	R50-01	50.72	51.22	JO02NF	Stowmarket	110.9Hz	G7CIY	1
GB3GC	R50-02	50.73	51.23	IO70WN	Gunnislake	77.0Hz	M0YDW	1
GB3SL	R50-02	50.73	51.23	IO75XX	Kilsyth	103.5Hz	GM4COX	1
GB3XD	R50-02	50.73	51.23	IO93WH	Louth	71.9Hz	G7AJP	1
GB3UM	R50-03	50.74	51.24	IO92IQ	Markfield	77.0Hz	M1NAS	1
GB3LP	R50-04	50.75	51.25	IO83MK	Liverpool	77.0Hz	M1SWB	3
GB3TQ	R50-04	50.75	51.25	IO80FK	Paignton	77.0Hz	G0AZX	1
GB3HF	R50-05	50.76	51.26	IO93HO	Barnsley	71.9Hz	G4LUE	1
GB3DB	R50-06	50.77	51.27	JO01HR	Danbury	110.9Hz	G6JYB	2
GB3FH	R50-06	50.77	51.27	IO81OH	Somerset	77.0Hz	G4RKY	1
GB3PX	R50-07	50.78	51.28	IO92XA	Barkway	77.0Hz	G4NBS	1
GB3TY	R50-07	50.78	51.28	IO74BS	Carrickfergus	110.9Hz	GI0PCU	2
GB3SX	R50-08	50.79	51.29	IO93BA	Stoke on Trent	103.5Hz	G4SCY	1
GB3WY	R50-09	50.8	51.3	IO93EP	Wakefield	82.5Hz	G1XCC	1
GB3ZY	R50-09	50.8	51.3	IO81QJ	Bristol	77.0Hz	G4RKY	1
GB3FX	R50-10	50.81	51.31	IO91OF	Farnham	82.5Hz	G4EPX	1
GB3ZW	R50-10	50.81	51.31	IO82HL	Newtown	103.5Hz	GW4NQJ	4
GB3HM	R50-11	50.82	51.32	IO93GA	Belper	71.9Hz	G8IQP	1
GB3GT	R50-12	50.83	51.33	IO82QJ	Clee Hill	103.5Hz	G1MAW	1
GB3JX	R50-12	50.83	51.33	JO02QP	Norwich		M0ZAH	1
GB3AM	R50-13	50.84	51.34	IO91QP	Amersham	77.0Hz	G0RDI	1
GB3CT	R50-14	50.85	51.35	IO92MG	Northampton	77.0Hz	G7SYT	1
GB3PD	R50-14	50.85	51.35	IO90KT	Portsmouth	71.9Hz	G8UCY	2
GB3VI	R50-15	50.86	51.36	IO92BL	Birmingham	67.0Hz	G8NDT	1

Notes:
1: Operational; 2. Not operational; 3. Poor sensitivity; 4. Cleared to go; 5. Cross-band repeater, 10m to 6m

APPENDIX D – IARU REGION 1
4M BEACONS (60 – 71MHZ)
(Source: IARU Region 1 beacon list compiled by G0RDI, RSGB website).

Freq MHz	Callsign	Nearest town	Locator	Antenna	Heading	Pwr (W)	Status	Latest update
60.05	GB3RAL		IO91EN				QRV	06/08 G6JYB
70.00	GB3BUX	Buxton, Derbys	IO93BF	2xTurnstile	Omni	20		06/02 G3UUT
70.007	GB3WSX	Yeovil	IO80QW	Yagi	70°	150		06/05 G0RDI
70.01	?						TEP Studies	02/04 DL8HCZ
70.01	J5FOUR		IK21EV	4el Yagi	020°	20	QRV	03/08 CT1FFU
70.012	OX4M		GP15EO	Dipole	Omni	25		11/05 OZ2TG
70.014	S55ZRS	Mt Kum	JN76MC				Planned	06/98 S57C
70.015	ZR6FOR						Planned	09/99 ZS5JF
70.016	GB3BAA	Nr Tring	IO91PS	2xDipole	Omni	25		01/09 G0RDI
70.018	OH2FOUR	Lohja	KP20DH	Dipole	Omni	15		07/11 OH6DD
70.02	GB3ANG	Dundee	IO86MN	3 el Yagi	160°	100		06/02 G3UUT
70.021	OZ7IGY	Jystrup	JO55WM	Big Wheel	Omni	40		08/06 OZ7IS
70.023	PI7EPO	Rijswijk	JO22EA	Halo	Omni	20	QRT	10/09 PA0EZ
70.025	GB3MCB	St Austell	IO70OJ	2 el Yagi	45°	40	QRV	04/09 G6JYB
70.027	GB3CFG	Carrickfergus	IO74CR	2x3 el Yagi	45°/135°	65	QRV	05/09 G0HIK
70.029	S55ZMB		JN76VK	4 el Yagi	310°	5		09/99
70.03	S56A		JN76GB			1	Personal	06/98 G3NKS
70.03		*G Personal Beacons*						
70.033	OH5RBG	Kouvla	KP30HV	7dBi	232°	15		07/11 OH6DD
70.035	OY6BEC		IP62OA		225°	25		11/05 OZ2TG
70.04	SV1FOUR	Athens	KM17UX	Halo		10	QRV	10/06 OD5TE
70.045	OE5QL		JN78CJ	Vertical	Omni			04/09 OE5MPL
70.05		*Proposed for IARU IBP*						
70.05	GB3RAL		IO91EN			15	QRV	06/08 G6JYB
70.06	HG1BVC	Hörman-forrás	JN87FI	2el Yagi	315°	10		03/13 HA5NF
70.063	LA2VHF		JP53EG	4 el	015°	25	PLANNED	01/10 LA0BY
70.065	LA5VHF		JP48AD	GP	Omni	20	PLANNED	11/10 LA0BY
70.081	HG8BVC	Kisráta	KN06HT	X Dipoles	Omni	5		03/13 HA5NF
70.081	LA7VHF		JP99LQ	4 el	190°	40	QRV	06/10 LA0BY
70.085	A92C/B	Mina Salman	LL56HE	Dipole	Omni	10		06/12 A92IO
70.109	IZ1DYE/B	Giaveno	JN35PA	Dipole	Omni	5		04/13 IK1YWB
70.114	5B4CY	Zyghi	KM64PR	4 el Yagi	315°	15		06/98 PA3BFM
70.117	OK0EE	Bystrice n Pernšt	JN89CK	X Dipoles	Omni	10	QRV	07/10 OK1HH
70.13	EI4RF	Dublin	IO63WD	2x5 el Yagi	45°/135°	25		08/95 G3NKS

Note: If the 'Status' column for a beacon is blank the beacon can be assumed to be 'on air'. The reliability of this information can be judged by the source and date in the 'last update' column.

APPENDIX E – IARU REGION 1
6M BEACONS (30 – 60MHZ)
(Source: IARU Region 1 beacon list compiled by G0RDI, RSGB website).

Freq MHz	Callsign	Nearest town	Locator	Antenna	Heading	Pwr (W)	Status	Latest update
40.021	OZ7IGY	Jystrup	JO55WM	Dipole	75°/255°	22	QRV	07/07 OZ7IS
40.05	GB3RAL		IO91EN				QRV	06/08 G6JYB
50.0	GB3BUX	Buxton	IO93BF	Turnstile	Omni	20		06/03 G4IHO
50.0	9A1CAL		JN86EL	Turnstile	Omni	1	QRT 06?	04/13 G0RDI
50.001	IW3FZQ/B	Monselice PD	JN55VF	5/8 Vert	Omni	8		04/13 IK1YWB
50.004	I0JX/B	Roma	JN61HV	5/8 Vert	Omni	5		04/13 IK1YWB
50.0047	4N0SIX	Belgrade	KN04FU	Dipole	Omni	1		
50.007	HG1BVB	Hörman-forrás	JN87FI	X Dipoles	Omni	20		03/13 HA5NF
50.008	EA3RCC	Barcelona	JN11BP	GP	Omni	5	QRV	09/07 EA3ABN
50.008	I5MXX/B	Pieve a Nievole	JN53JU	5/8 Vert	Omni			04/13 IK1YWB
50.0108	SV9SIX	Iraklio	KM25NH	Vert dipole	Omni	30		05/02 G3USF
50.011	OK0EK	Kromeriz	JN89QG	2xDipole	Omni	10	QRT	11/12 OK1HH
50.012	OH1SIX	Ikaalinen	KP11QU				PLANNED	07/11 OH6DD
50.012	OX3SIX	Kulesuk	GP15EO	Vert Dipole	Omni	100		11/05 OZ2TG
50.013	LZ1JH		KN22TK	GP	Omni	1		11/00 LZ2HM
50.013	CU3URA	Terceira, Azores	HM68	5/8 Vert		5		06/98 EA7KW
50.014	OK0SIX	Nr Milevsko	JN79FM	Vertical	Omni	2.5	* 50.414	11/12 OK1HH
50.016	GB3BAA	Nr Tring	IO91PS	Vert Dipole	Omni	10	* 50.416	02/10 G0RDI
50.017	OH0SIX	Stalsby	JP90XI	Dipole	0°/180°	3		07/11 OH6DD
50.019	IZ1EPM/B	Saronsella TO	JN35WD	5/8 Vert	Omni	10		04/13 IK1YWB
50.02	IW8RSB/B	Simeri CZ	JN88HW	5/8 Vert	Omni	5		04/13 IK1YWB
50.02	CS5BSIX		IM58is	Yagi	Transatlantic	5	PLANNED	07/08 CT1END
50.0207	IK5ZUL/B	Follonica GR	JN52JW		Omni			04/13 IK1YWB
50.021	OZ7IGY	Jystrup	JO55WM	Big Wheel	Omni	48	* 50.471	03/10 OZ7IS
50.022	S55ZRS	Mt Kum	JN76MC	Gnd Plane	Omni	1	* 50.422	04/13 G0RDI
50.022	HG8BVB	Gerla	KN06OQ	Gnd Plane	Omni	5		03/13 HA5NF
50.023	LX0SIX	Bourscheid	JN39BF	Hor Dipole	0°/180°	10		05/02 G3USF
50.023	SR5FHX	Slubianka	KO02LL	3el Yagi	240°	3		04/05 SP6LB
50.0247	UN1SIX	Kazakhstan	MN83KE	Gnd Plane	Omni	12		05/02 G3USF
50.025	9H1SIX	Attard	JM75FV	Gnd Plane	Omni	7		
50.025	OH2SIX	Lohja	KP20DH	Dipole	Omni	50		07/11 OH6DD
50.026	IQ4FA/B	Ferrara	JN54TU	Gnd Plane	Omni	5		04/13 IK1YWB
50.026	SR9FHA	Chorgawica	KN09AS	5/8 Vert	Omni	5		04/07 SP6LB
50.027	CN8LI	Rabat	IM64	J-Pole	Omni	8		03/99 CN8LI
50.027	SK7SIX	Hultsfred	JO77UM		Omni	15		12/01 SM0KAK
50.0276	SR6SIX	Sztobno / Wolow	JO81HH	Gnd Plane	Omni	10		07/94 SP6LB
50.028	5T5SIX		IK28AC	Gnd Plane	Omni	50	Proposed	03/04 5T5SN
50.028	SR8SIX	Sanok	KN19CN				Non op	05/02 G3USF
50.028	SR3FHB	Chelmce	JO91CQ	Dipole	Omni	5		04/07 SP6LB
50.031	HG7BVA	Gyömro	JN97QJ	Gnd Plane	Omni	5		03/13 HA5NF
50.0315	CT0SIX	Tavarde	IM59QM	Hor Dipole	90°/270°			09/06 CT2IRJ
50.033	OH5RAC	Kuusankoski	KP30HV	2dBd	200°	20		08/06 OH6DD

Freq MHz	Callsign	Nearest town	Locator	Antenna	Heading	Pwr (W)	Status	Latest update
50.033	OH5SHF	Kouvola	KP30HV	2dBd	180°	20		07/11 OH6DD
50.035	OY6SMC	Faroe Is	IP62MB					11/05 OZ2TG
50.035	CQ3SIX	Madeira Is	IM21IP	Halo	Omni	10		05/07 G0LGS
50.0366	SR2SIX	Bydgoszcz	JO93BC					05/02 G3USF
50.037	CT1ART	Serra d Caldeirao	IM67AH	6el Yagi		40		01/08 CT1HZE
50.0375	ES0SIX	Hiiuma Island	KO18CW	Hor dipole	90°/270°	15		12/01 SM0KAK
50.04	SV1SIX	Athens	KM17UX	Vert Dipole	Omni	30		08/00 G3UUT
50.041	ON0SIX	Vieux Genappe	JO20EP			50		05/07 ON4PC
50.0424	GB3MCB	St Austell	IO70QJ	Dipolo	90°/270°	40		04/09 G6JYB
50.044	ZS6TWB	Haenertsburg	KG46XA	5/8 Vert	Omni	15		11/00 ZS6PJS
50.045	OX3VHF	Qaqortoq	GP60XR	Gnd Plane	Omni	20		11/05 OZ2TG
50.045	SR2FHM	Gdansk	JO94HI	Dipole		7		04/07 SP6LB
50.046	JW5SIX	Hopen	KQ26MM	Vert Dipole	Omni	10	* 50.445	02/10 LA0BY
50.047	JW7SIX	Isfjord Radio	JQ68TB	3 el Yagi	180°	30	* 50.447	02/10 LA0BY
50.047	YO2S	Timisoara	KN05PS	Dipole		1		06/99 YO2IS
50.0472	4N1SIX	Belgrade	KN04OO	Vee	Omni	10	Non op	05/02 G3USF
50.0485	TR0A	Libreville	JJ40	5 el Yagi	0°	15		05/02 G3USF
50.049	JW9SIX	Bjornoya	JQ94LM	Dipole	Omni	10	* 50.449	02/10 LA0BY
50.05	GB3RAL		IO91IN			17	QRV	06/08 G6JYB
50.05	IARU IBP							06/02 G3UUT
50.051	LA7SIX	Malselv	JP99EC	4 el Yagi	190°	100	* 50.451	10/02 LA0BY
50.052	EI?????						Planning	06/05 G0RDI
50.052	SK2CP	Kiruna	KP07MV		Omni	30		05/10 SM2UHF
50.053	PI7SIX	Utrecht	JO22NC	Dipole	0°/180°	12		07/03 PE2KP
50.054	OZ6VHF	Oestervraa	JO57EI	Turnstile	Omni	25		11/05 OZ2TG
50.057	TF3SIX	Reykjavik	HP94BC		Omni	20		12/07 OZ7IS
50.057	IT9X/B	Messina	JM78SG	Loop	Omni	10		04/13 IK1YWB
50.058	CU?????						Planning	06/05 G0RDI
50.058	IW0DTK/B	Latina	JN61TG	5/8 Vert	Omni	10		04/13 IK1YWB
50.058	HB9SIX	Nr St Gall	JN47QF	J-pole	Omni	6		07/08 HB9RUZ
50.058	IQ4AD/B	Parma	JN54DT	5/8 Vert	Omni			04/13 IK1YWB
50.06	GB3RMK	Inverness	IO77UO	Dipole	0°/180°	32		08/09 G6JYB
50.061	EA3VHF	Gerona	JN11MV	Vertical	Omni1		Intermittent	05/02 G3USF
50.062	OZ2VHF	Esbjerg	JO45FL	Hor Dipole	0°/180°	1		06/99 G3USF
50.0625	GB3NGI	Ballymena	IO65VB	Halo	Omni	30	QRV	06/10 G6JYB
50.063	LY0SIX		KO24PS	6 el Yagi	250°	7		03/04 LY2MW
50.064	GB3LER	Lerwick	IP90JD	Dipole	0°/180°	30	QRV	04/09 G6JYB
50.065	GB3IOJ	Jersey	IN89VE	Halo	Omni	25	QRV	01/13 G6JYB
50.066	OE3XAC	Kaiserkogel	JN78SB	5/8 Vert	Omni	10	QRV	01/10 OE1MCU
50.067	OH9SIX	Pirttikoski	KP36OI	2 X-dipole	Omni	35		07/11 OH6DD
50.068	HG8BVD	Kisráta	KN06HT	Gnd Plane	Omni	5		03/13 HA5NF
50.07	SK3SIX	Edsbyn	JP71XF	X Dipole	Omni	10		07/98 SM6CEN
50.072	SM0???		JO99				PLANNING	05/09 SM0TSC
50.0735	EA8SIX		IL28GD		Omni	14		05/02 G3USF
50.0745	ED7YAD	Malaga	IM76qo	Loop	Omni	15	QRV	03/09 EA7UU
50.075	YO3KWJ		KN34BJ	Gnd Plane	Omni	5		05/99 YO3JW
50.076	CS1RLA/B	Aldeia de Chaos	IM57PX	Dipole	Omni	2.5		07/07 CT1FBF

Freq MHz	Callsign	Nearest town	Locator	Antenna	Heading	Pwr (W)	Status	Latest update
50.076	CS5BLA	Aldeia de Chaos	IM57px	Horiz		2.5	QRV	04/09 CT1FBF
50.077	DL????						PLANNED	01/01 DJ3TF
50.078	OD5SIX	Lebanon	KM74WK	1/4 Vert	Omni	7		01/96 OD5SB
50.079	JX7SIX	Jan Mayen	IQ50RX	3 el Yagi	160°	40	* 50.479	02/10 LA0BY
50.08	4X4SIX	Tel Aviv	KN72JB	Dipole		3		05/02 G3USF
50.08	UU5SIX	Nr Yalta	KN74AL	Dipole		10		05/02 G3USF
50.084	UT5G	Petri	KN66LS	Gnd plane	Omni	10		04/00 UR4LL
50.0847	UR4LL		KO70XG	H dipole	Omni	8		04/00 UR4LL
50.315	FX4SIX	Neuville	JN06CQ	2xdipole	Omni	25	* 50.448	04/13 G0RDI
50.32	F8BHU	Nevers	JN17NA	3 el Yagi	315°	5	* 50.434	04/13 G0RDI
50.412	C30SIX	Engolasters	JN02SM	Loop	Omni	40	PLANNING	02/10 EA7KW
50.415	IW3ICH/B	Rovigo	JN55WD	I-Pole	Omni	1		04/13 IK1YWB
50.445	EA1WD	Sena/Limpias	IN83JJ	Big Wheel	Omni5		In progress	01/13 EA1DPP
50.457	IW0DAQ/B	Albano Laziale	JN61HQ	Quad Loop	Omni	2		04/13 IK1YWB
50.459	LA9SIX		JP50EV	5/8 Vert	Omni	25	PLANNING	11/12 LA0BY
50.471	OZ7IGY	Jystrup	JO55WM	Big Wheel	Omni	48	PLANNED	03/10 OZ7IS
50.473	OK0EQ	Namest n/O	JN89BE	Dipole	Omni	4	PLANNED	11/12 OK1HH
50.475	OK0NCC	Praha	JN79EW	Dipoles	Omni	5	PLANNED	11/12 OK1HH
50.477	OK0EXD	Ricany	JN79IW	Vertical	Omni	10	PLANNED	11/12 OK1HH
50.4995	5B4CY	Zyghi	KM64PR	Gnd Plane	Omni	20		07/92 5B4JE
50.5203	SZ2DF		KM25	4x16 el	30°/330°	1000		05/02 G3USF
52.45	VK5VF	Mt Lofty Adelaide	PF95	Turnstile	Omni	10		05/02 G3USF

Notes:

1. Note: At an Interim C5 Meeting (Vienna, 02/2010) it was proposed that the 6m beacon band will be relocated to 50.4000 – 50.5000MHz. As a general principle, all beacons should move to the planned frequencies as soon as possible. An asterisk (*) in the 'Status' column indicates that the beacon will move to the frequency indicated.

2. If the 'Status' column for a beacon is blank the beacon can be assumed to be 'on air'. The reliability of this information can be judged by the source and date in the 'last update' column.

3. A world-wide list of 6m beacons compiled by Martin Harrison, G3USF, for the RSGB Propagation Studies Committee may be downloaded from www.keele.ac.uk/depts/por/50.htm

APPENDIX F – LOCATORS

The IARU Locator System, usually just called 'Locator', provides a means of pinpointing stations throughout the world. It is most often used by operators on 50MHz and above as a means of calculating the distance between two stations (for use by operators on the upper microwave bands, it can have eight digits, though only the first six are dealt with here). The system is based upon latitude and longitude.

There are three sizes of 'rectangle'. The largest, known as a 'field', is 20° of longitude (east-west) by 10° latitude (north-south), and is designated by two letters. Most of Britain is in IO field, as shown below.

The next rectangle, known as a 'square' (though it is actually neither truly square nor rectangular!) is 2° of longitude by 1° of latitude. One hundred squares make up one field and these are given numbers 00 in the south-west corner to 99 in the north-east, e.g. Dublin is in IO63. Finally, each square is divided into 576 'sub-squares', 5 minutes of longitude by 2.5 minutes of latitude, and given letters from AA to XX.

To find out your locator, first use a map of your area to determine your exact latitude and longitude, then use the map and the diagrams opposite to pinpoint your locator. Computer programs and on-line calculators are available to do this more easily: on-line Lat/Long to/from Locator calculators are at www.arrl.org/locate/grid.html and www.amsat.org/cgi-bin/gridconv while an on-line National Grid Reference (NGR) to Locator calculator is at www.ntay.com/contest/NGR2Loc.html

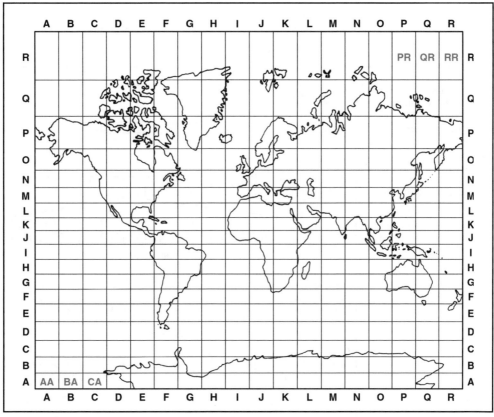

The IARU Locator system may be used throughout the world without repeats. The map above shows the fields that make up the first two letters of the Locator. Examples are shown at two of the corners.

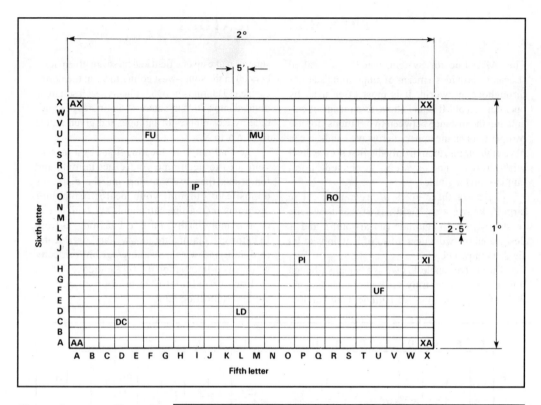

Above: A square (the numbered part of the Locator) is divided into 576 sub-squares, designated AA to XX. Each sub-square is 5' W–E and 2.5' N–S.

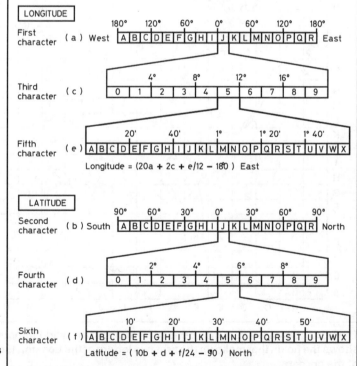

Right: The final two letters may be calculated thus.

Index